God Did It

Betty J. White

PublishAmerica
Baltimore

© 2004 by Betty J. White.
All rights reserved. No part of this book may be reproduced, stored in a retrieval system or transmitted in any form or by any means without the prior written permission of the publishers, except by a reviewer who may quote brief passages in a review to be printed in a newspaper, magazine or journal.

First printing

ISBN: 1-4137-3550-9
PUBLISHED BY PUBLISHAMERICA, LLLP
www.publishamerica.com
Baltimore

Printed in the United States of America

Dedication

To my beloved husband, Lawrence,
and precious children.
Larry
Rhonda
Steven
Sherry
Cynthia
Beverly
Frank
John
Mary
William
James
Charles

God Did It

Prologue

An exegesis of the first seven chapters of Genesis of the *Holy Bible*[1] was written to prove that "evolution" was a creative process wrought by God.

Verses from *The Authorized King James Version* are provided for convenience. This version is so worded as to reveal certain concepts that may not be recognizable in some other versions.

Scientific experiments have revealed that living organisms come only from other living organisms. The only exception to this truth, of course, was when God created living organisms from what apparently was nonliving matter in the third day of creation. There was, undoubtedly, much chemical evolution happening during the first two days of creation. Therefore, God having been the Creator, nullifies the belief that *spontaneous generation*[2] has occurred since the third day of creation. Genesis shows that there was an unfolding in creation/evolution by depicting the rising of life from simple species to complex ones: God began the third day by creating single-celled organisms and ending that day with vegetation. The fourth day life became more complex; the fifth day even more so. The sixth day, God brought forth man—the most complex of all natural life.

The geological record and dating apparatus tell us that life began on earth billions of years ago, but the Bible tells us that Adam and Eve lived on earth about six thousand years ago. Truth is an absolute, so there cannot be gaps between scientific facts and Biblical truths. Therefore, some interpretations of the word of God and how He created must be revamped.

The Bible and the universe speak loudly of the "order and consistency" of God's creative work: "For I am the Lord, I change not ..." (Malachi 3:6). God would be a God of "change" if He had created Adam from dust into a fully-grown man. Adam was born and experienced consistent aging like all men, because the Bible says that he died when he was nine hundred and thirty years old. It would have been disorder and inconsistency if God had begun Adam's life when he was a fully-grown man, because no other earthly life has begun that way. To believe that Adam had not been born like other men, but created instantaneously, with a navel, would also be disorder and a "lie," as a navel proves that Adam was born from a preceding life. God cannot lie! To believe that Adam was born or created without a navel would be inconsistent and very odd, as all men are made in the image of God and have navels. Therefore, our unchanging God did not create any life instantaneously, or in a literal twenty-four hour period of time, but created all life from dust by an orderly and consistent process of evolution, which took several billion years.

Mutations, natural selection, adaptations, and any other changes that occurred during the evolution of species were all part of God's sublime handiwork. We do not need to know all the ramifications of evolution to appreciate the concept that God created natural life by an unfolding process. There were no breaks or lulls in God's evolutionary work. From the first single-celled organisms until man was made, there was a continuous evolving/unfolding process from simple to complex, as certain lives gave rise to new species.

Evolution is the way God created! God took many years to create all life. There were transitional stages to different kinds of life, for God caused some of those lives to evolve, and as time passed, life became more complex. His long-term goal was to make man in His "image and likeness."

Evolution has ended! All life has evolved to its complex end. That means that man has evolved into the *image* and *likeness* of God. The Bible is a safe guide to use when scrutinizing scientific theory. One theory that the Bible discredits is belief that evolution is still happening. To exploit that belief, without proof, is as wrong as being stiff-necked about changing some traditional beliefs that contradict Biblical facts.

Many have rejected the process of evolution as a creative means, because evolution has been mistakenly presented as a "random chance" happening. It can unquestionably be said that God the Creator cannot be disproved, so random chance evolution cannot be proven true! God was actually the overseer in what appeared to be random chance in nature.

Many have asked: "Which came first, the chicken or the egg?" The transitional stages of making a chicken with very subtle changes, over long periods of time, could only be done by the Creator. One day an "almost finished" chicken laid an egg with a "finished and fixed" chicken inside. God did not say that certain lives had to be perfectly formed "after his kind" (Genesis 1:24) until the species were finished and fixed. Therefore, the egg came first!

Charles Darwin said: "It is utter nonsense to think presently about the origin of life; one might just as well contemplate the origin of matter."³ Darwin was able to read the many signs of evolution left in the earth, but he was unable to understand how matter originated, even though he was aware that there had to be an event. In Darwin's day, no one had heard of the Big Bang.⁴ Such a cataclysmic event was the "origin of matter" that occurred in the beginning as the first act of God's creative will! Cosmologists were able to realize the Big Bang when they discovered that the universe has been expanding. If one could have filmed the expansion for billions of years and would have run the film backwards; one would have been able to see all matter move towards a center of origin.

Organic life was made from matter that originated during the first creative act. The materialistic existence of everything in the universe is made from that matter, which constitutes a certain number of elements. Although our natural bodies and the stars contain like matter; natural life was made from the dust of the earth, and not from the dust of any other planet. This fact is verified when God said to Adam:

> In the sweat of thy face shalt thou eat bread, till thou return unto the ground; for out of it wast thou taken: for dust thou art, and unto dust shalt thou return (Genesis 3:19).

God having created life by an evolving process is not a refutation of the fact that man was made from dust. God made the first cells from the dust (elements) of the earth, and life continued to evolve by His power throughout the six creative days towards the ultimate creation called "man."

In the word of God we also read: "… that one day is with the Lord as a thousand years, and a thousand years as one day" (II Peter 3:8). With this Scripture in mind, we can speculate on how God's creative days lasted for billions of years. Scientists have stated that the Big Bang occurred about 20 billion years ago. Whether it was 20 billion years ago or twice that amount is irrelevant, for God's creative days were from the origin of matter, until God made man in His image.

Scientists have tried in a vain attempt to bring about a form of life by mixing together some inorganic substances. They have acquired the ability to develop *proteinoids*[5] by the polymerization of amino acids, but man will never be able to develop any form of life. Such a conquest would require an indeterminable amount of ingenuity and cosmic time, which are attributes that belong only to God.

Every possible form of life has been made because there is a "fullness" of creation! God's creative work would have been incomplete if He had left a possible form of life in a void of nonexistence. Before a new form of life could be made, there would first have to be a new element created. Since all elements were created in the beginning, then there is no way that a new form of life could develop. If God gave man the ingenuity and time, man could not create a different matter, animate or inanimate, from that which is or has been created:

> The thing that hath been, it is that which shall be; and that which is done is that which shall be done: and there is no new thing under the sun (Ecclesiastes 1:9).

During the six creative days, God undoubtedly caused some mutations to occur as He was creating new species. Those kinds of mutations no longer occur, as creation is over and life no longer evolves! The only kinds of mutations that have occurred since then were those caused by accidents, diseases, and manipulations by man. Genetic engineering, such as gene splicing, is a manipulation, which causes mutations, not creations. Gene splicing is used to alter the hereditary apparatus of living organisms, which may be beneficial to some unhealthy ones. It may also change the behavior of certain ones to the benefit of mankind, but manipulation does not benefit healthy organisms, as God's creative work cannot be improved upon.

All species have become specialized (finished and fixed), each one on one of the six days of creation. However, many extinct creatures that existed during the creative days were only transitional stages of life, which proves the "entropy law" was built into nature, because all matter will eventually return to dust.

Life continued on after God stopped the evolutionary process, but only by species reproducing "after his kind" or by "hybridization." In 1979, an example of hybridization was the ape that was born in captivity at Atlanta's Grant Park Zoo. She was dubbed a *siabon*[6], the result of the mating of two species of apes—a male gibbon and a female siamang. God allows man to be a helping cause of these two kinds of reproduction. Hybridizing is not creating; it is only crossing that which is already created. Man's abilities are limited when it comes to crossing species because in a genetic sense, many are too far apart. Man and ape are too far apart to be crossed. They have similar characteristics because both species evolved on different paths from a common ancestor. However, the offspring of most species that have been crossed are sterile, which is God's way of protecting the identity of species.

Fossils of prehistoric primates, animals, birds, insects, fish, plants, and marine invertebrates cannot be denied. They are signs left in the earth, by the will of God, for our enlightenment of His creation! The marine invertebrates were early lives from the Cambrian period, of the Paleozoic era, dating back about 550 million years ago.

Since God made the first cells from the dust of the earth, then it is impossible that organic life could have originated on any other planet. Life had already developed and advanced to the vegetative stage on earth, before God had placed any other planet in the firmament. There is a theory called *panspermia*[7] that states that a highly evolved life from another solar system may have planted life seeds on our earth. Pure logic tells us that a life form from somewhere else would also have need of a creator or seed planter. God would only be repeating His creative work if He started life from the dust of another planet, because there is already a "fullness" of creation.

The word of God reveals what He created in six cosmic days. The correct interpretation of the word of God reveals: (a) Organic life could not have been created anywhere, before the creation of matter. (b) Organic life could not have been created anywhere but on planet earth before the fourth day of creation because there were no other planets before that time. (c) God has not created anything since He made man on the sixth day of creation, somewhere between six thousand and fifty thousand years ago, because His creative work ended after they (male and female) were made. It took four cosmic days for all life to appear on planet earth before God stopped creating. God would have needed two favorable cosmic days left for chemical evolution to occur on another planet, and then it would have been time to rest from His creative work. Therefore, God's revelation to us that natural life was made only from the dust of the earth cannot be contradicted. Since the word of God is perfect evidence of the Creator, then any theory or doctrine that deviated from the word of God was misconstrued from the start.

Some UFO (unidentified flying object) enthusiasts are hoping that a super intelligent life (that is, outside of God) may come and save our so-called "failing" planet. Their hopes are as illusive as those strange crafts. They readily appear and disappear. There have been reports of human contacts with weird, unearthly beings, but from all indications, none were here on a mission of mercy. In fact, there have been some very gruesome acts attributed to them. Since the fullness of creation has been made manifest, and man being the zenith of earth's manifestations, then there must be transcendent powers in operation: "And no marvel; for Satan himself is transformed into an angel of light" (II Corinthians 11:14). There is not enough credit given to Satan and his followers for the power they possess. The word of God says that Satan is: "… the prince of the power of the air, the spirit that now worketh in the children of disobedience" (Ephesians 2:2). From his kingdom come apparitions, false prophets, false teachers, occults, atheism, possessions, oppressions, obsessions, superstitions, and a host of other ills. The only way to overcome the power of Satan and his followers is with the omnipotent power of God: "Take heed to yourselves, that your heart be not deceived, and ye turn aside, and serve other gods, and worship them" (Deuteronomy 11:16). The existence of transcendent powers cannot be proven in a test tube. They are spiritual, and God has revealed their existence to us in His word: "Sanctify them through thy truth: thy word is truth" (John 17:17).

The extent of the six days of creation have been made manifest in the fullness of God's foreordained will, and an interpretation of how He went about creating all things follows.

Genesis

Chapter 1

1. In the beginning God created the heaven and the earth.
2. And the earth was without form, and void; and darkness was upon the face of the deep. And the Spirit of God moved upon the face of the waters.
3. And God said, "Let there be light": and there was light.
4. And God saw the light, that it was good: and God divided the light from the darkness.
5. And God called the light Day, and the darkness he called Night. And the evening and the morning were the first day.

"In the beginning" tells something about time. The time related here was the event that existed between (1) the beginning, which was the start of God's creative work, and (2) the end, which was the cessation of God's creative work. In the beginning must entail the full six creative days, as the earth was not made manifest until the third day (Genesis 1:9,10). Since the third day alone or the first three days altogether cannot be marked as the "beginning," then the manifestation of the fullness of creation, which took six full cosmic days must be included. All six cosmic days must be considered the "beginning and end" of creation! All the rest of time is the seventh day or day of rest.

God gave man the ability to comprehend time, because man would live in a time dimension. Man cannot totally comprehend timelessness or eternity, until he lives in a timeless or eternal dimension. The soul-filled man is the only life on earth that can and will experience both time and timeless dimensions, because he has two bodies: "It is sown a natural body, it is raised a spiritual body. There is a natural body, and there is a spiritual body" (I Corinthians 15:44).

Man's natural body has a brain or mind, and like other animals, it guides the functions of the natural body. Unlike other animals, however, man made in the "likeness" of God has a spiritual body that has a brain or mind that is the "soul" or "subconscious mind." Therefore, the soul is part of the spiritual body, but the spiritual body is not part of the soul for the spiritual body is the whole. In the same way it can be said that the natural brain is part of the natural body, but the natural body is not part of the natural brain for the natural body is the whole.

When the natural brain is incapacitated or dead, it cannot function in any realm. A person has awareness in his spiritual body, after his two bodies separate. The spiritual body, which is the "inner man" and sometimes referred to as the "heart," simply will not leave the natural body permanently, until the natural body is finally dead, for God made the two to be inseparable until death. Some people have claimed that their spiritual body left their natural body for a short time. Since their natural body did not permanently die, then their spiritual body reentered their natural body and they were able to tell of the experience. Therefore, it is evident that the spiritual brain works conjunctively with the natural brain, in the sense that without the natural brain, the spiritual brain cannot manifest its thoughts into the natural realm.

"God created." There are no other deities who can create:

> For by him were all things created, that are in heaven, and that are in earth, visible and invisible, whether they be thrones, or dominions, or principalities or powers: all things were created by him, and for him: And he is before all things, and by him all things consist (Colossians 1:16,17).

There was only natural darkness in the deep void that now holds planet earth. When God spoke the words: "Let there be light" (that was when the Big Bang occurred) the abstract, ethereal, and materialistic light of God became manifest. Therefore, the first light was not the natural light of the sun, moon, and stars, for they were created on the fourth day. Created within the realm of that first light: (1) inorganic matter, (2) the laws of the universe, (3) the nature of opposites, and (4) heavenly entities. Some of the universal laws are light, heat, color, motion, sound, and gravitation. The nature of opposites tell us that God created spiritual darkness as a reciprocal to spiritual light, for without opposites it would be impossible to have a true concept of abstract things. Two Biblical verses that prove the nature of opposites and that God created both spiritual and natural light and darkness:

> Yea, the *darkness* hideth not from thee; but the night shineth as the day: the *darkness* and the *light* are both alike to thee (Psalms 139:12).

But if thine eye be evil, thy whole body shall be full of *darkness*. If therefore the *light* that is in thee be *darkness,* how great is that *darkness* (Matthew 6:23)!

God saw and was pleased with the light He created, and He set down a standard of separation of the created nature of opposites. Two kingdoms represent the broadest sense of that separation: (1) the kingdom of the light of God that is good and positive, and (2) the kingdom of the darkness of Satan that is evil and negative. The created light of God proves His goodness in a positive way. Although God has freedom of will, He cannot deviate from His goodness, as He is representative of the very light He created. God is "light" in the absolute, just as Satan is "darkness" in the absolute!

God created all things in six cosmic days! Heavenly entities were created during the first cosmic day when spiritual light was being made manifest. Of Satan, God said: "Thou wast perfect in thy ways from the day that thou wast created, till iniquity was found in thee" (Ezekiel 28:15). God created spiritual entities as free agents, but He certainly did not will for anyone to partake of the Dark Kingdom. The traitorous act of Satan and his followers did not occur in the heavenly realm of God's light. The act became manifest when the stepping over into spiritual darkness occurred:

> And the angels which kept not their first estate, but left their own habitation, he hath reserved in everlasting chains under darkness unto the judgment of the great day (Jude 1:6).

Satan is the "prince of devils" (Matthew 12:24) because he was the first to step over into the spiritually dark realm. He and his followers cannot repent; neither can they be redeemed. They chose the dark side and can never deviate from it.

The "light and darkness," "Day and Night," and "evening and morning," are equivalents. The aspects of light and its opposites, with the foreordained literal days and nights, not yet distinguishable, constituted the first day of creation. That day was a long period of time, but certainly not a literal twenty-four hour day.

> 6. And God said, Let there be a firmament in the midst of the waters, and let it divide the waters from the waters.
> 7. And God made the firmament, and divided the waters which were under the firmament from the waters which were above the firmament: and it was so.
> 8. And God called the firmament Heaven. And the evening and the morning were the second day.

In the second creative day, God set down the firmament that is also called an "arch" or "vault," which the Bible depicts as a "closed universe." The arch or vault is what is understood as "space." God then divided the waters from above and below the space where planet earth would hang in the firmament. The firmament has three major divisions: (1) earth's surrounding atmosphere, (2) the outer space regions of galaxies, and (3) Paradise. God called these three divisions of space "Heaven" as did King Solomon who built a temple to the Lord:

> But will God indeed dwell on the earth? Behold, the *heaven and heaven of heavens* cannot contain

thee; how much less this house that I have builded (I Kings 8:27)?

Biblical references that speak of the third heaven as Paradise:

> I knew a man in Christ above fourteen years ago, (whether in the body, I cannot tell; or whether out of the body, I cannot tell: God knoweth) such a one caught up to the *third heaven*. And I knew such a man (whether in the body, or out of the body, I cannot tell: God knoweth). How that he was caught up into *paradise* and heard unspeakable words, which it is not lawful for a man to utter (II Corinthians 12:2-4).

Biblical references that depict the throne of God, as being above the three heavens:

> Be thou exalted, O God, *above* the heavens; let thy glory be above all the earth (Psalms 57:5).

> He that descended is the same also that ascended up far *above* all heavens, that he might fill all things (Ephesians 4:10).

> 9. And God said, Let the waters under the heaven be gathered together unto one place, and let the dry land appear: and it was so.
> 10. And God called the dry land Earth; and the gathering together of the waters called he seas: and God saw that it was good.

11. And God said, Let the earth bring forth grass, the herb yielding seed, and the fruit tree yielding fruit after his kind, whose seed is in itself, upon the earth: and it was so.
12. And the earth brought forth grass, and herb yielding seed after his kind, and the tree yielding fruit, whose seed was in itself, after his kind: and God saw that it was good.
13. And the evening and the morning were the third day.

In the third creative day, God separated the wet and dry matter of our planet. He called the land "Earth" and the waters "Seas." These two words are not only indicative of the earth's appearance in the firmament, but they also reveal it as God's chosen planet—the only place made for the creation of natural life. There were no other planets in the firmament during the third creative day to hold the earth in place by gravitational force, for Job said: "He strecheth out the north over the empty place, and hangeth the earth upon nothing" (Job 26:7).

God spoke and the earth brought forth vegetation in the third creative day, but the vegetation was the end result of that day of work—not the beginning. As we can well see, the word of God does not state all the details of creation. According to recent discoveries, life has existed on earth for about 4.5 billion years. The upstart of biological activity began after elements (dust) adhered to one another in a perfect formula of composition, heat, and time, until single-celled organisms developed. Those early organisms were asexual, as they reproduced by a process called "fission": The parent organism would reproduce by dividing into two or more parts, each becoming an independent individual. The earth was dark, warm, and moist, and the soupy seas served as incubators to the small, delicate organisms. They were nurtured in an environment that had only a small amount of free oxygen, for a higher level of oxygen would have killed them. Later on, *photosynthesizing organisms*[8] developed and increased the level of oxygen needed for the evolution of the more advanced organisms.

It had never rained upon the earth (Genesis 2:5). Without the lights of the sun, moon, and stars, moisture could not rise up high into the atmosphere. Water vapor hovered close to the ground and kept it wet enough for primitive vegetation of the seas to spread upon land. As time passed, variations of plant life increased and conditions of Genesis 1:12 prevailed. Mother earth had advanced to the "Devonian period, of the Paleozoic era," some 350 million years ago, which was about the time that God finished the third day of creation.

14. And God said, Let there be lights in the firmament of the heaven to divide the day from the night; and let them be for signs, and for seasons, and for days, and years:
15. And let them be for lights in the firmament of the heaven to give light upon the earth: and it was so.
16. And God made two great lights; the greater light to rule the day, and the lesser light to rule the night: he made the stars also.
17. And God set them in the firmament of the heaven to give light upon the earth,
18. And to rule over the day and over the night, and to divide the light from the darkness:

and God saw that it was good.
19. And the evening and the morning were the fourth day.

On the fourth creative day, God spoke into existence the sun, moon, and stars. The light from those heavenly bodies caused moisture to be drawn up high into the atmosphere, and for the first time caused rain to fall upon the earth. God placed the heavenly bodies in the firmament to give natural light upon the earth, but He did not put them there to be used as a habitation! Life in its natural form cannot exist on any other planet, because there are no other planets exactly like our earth:

> There is one glory of the sun, and another glory of the moon, and another glory of the stars: for one star differeth from another star in glory (I Corinthians 15:41).

The earth is the oldest planet in the universe! This fact may never be proven in a scientific way, but the Bible is a reliable source from which this truth is made known. Therefore, scientific inquiries should never be directed contrary to what the word of God reveals: Genesis reveals that evidence of the earth came about in the third creative day, and evidence of other planets came about in the fourth creative day. The components of them all were created at the same time, but their formations came about on a different creative day.

God's great power to create the planets proves His great power to govern them. The proof of God's governing wipes out the idea of "random chance" evolution. There is a consistent order of the planets that enables us to mark time. That order could not have been broken by random chance the way that it was broken by God, on two specific occasions:

(1) Then spake Joshua to the Lord in the day when the Lord delivered up the Amorites before the children of Israel, and he said in the sight of Israel, Sun, stand thou still upon Gibeon; and thou, Moon, in the valley of Ajalon. And the sun stood still, and the moon stayed, until the people had avenged themselves upon their enemies. Is not this written in the book of Jasher? So the sun stood still in the midst of heaven, and hasted not to go down about a whole day (Joshua 10:12,13).

(2) And the Lord said unto Moses, Stretch out thine hand toward heaven, that there may be darkness over the land of Egypt, even darkness which may be felt. And Moses stretched forth his hand toward heaven; and there was a thick darkness in all the land of Egypt three days (Exodus 10:21,22).

The fourth day of creation was another long period of time. The literal days and nights became distinguishable, because the natural lights were in the firmament to divide the light from the darkness. At that time God had finished two-thirds of His six cosmic days of creative work.

20. And God said, Let the waters bring forth abundantly the moving creature that hath life, and fowl that may fly above the earth in the open firmament of heaven.
21. And God created great whales, and every

living creature that moveth, which the waters brought forth abundantly, after their kind, and every winged fowl after his kind: and God saw that it was good.
22. And God blessed them, saying, Be fruitful, and multiply, and fill the waters in the seas, and let fowl multiply in the earth.
23. And the evening and the morning were the fifth day.

In the fifth creative day life continued to evolve. God spoke and reptiles and fowls made their appearance. God revealed facts of evolution when He commanded the waters to bring forth creeping creatures and fowl life. The waters could not bring forth any kind of life unless it did so in an evolutionary way. What other conceivable way could water bring forth life? The creating/evolving or unfolding of life produced many transitional stages of life that were "after his kind," but after many years of slight variations of each kind, new species would emerge. The "protobird," which had both reptilian and bird characteristics, was a good example of a probable, transitional type of life. That creature developed in the "Jurassic period, of the Mesozoic era," about 155 million years ago, right after the great dinosaurs became dominant. On that creative day, God commanded lives to multiply. The seas filled with great whales and other sea life, and a variety of winged creatures developed and filled the earth. Another cosmic evening and morning passed as the fifth day of God's creative work ended.

24. And God said, Let the earth bring forth the living creature after his kind, cattle, and creeping thing, and beast of the earth

after his kind: and it was so.
25. And God made the beast of the earth after his kind, and cattle after their kind, and every thing that creepeth upon the earth after his kind: and God saw that it was good.
26. And God said, Let us make man in our image, after our likeness: and let them have dominion over the fish of the sea, and over the fowl of the air, and over the cattle, and over all the earth, and over every creeping thing that creepeth upon the earth.
27. So God created man in his own image, in the image of God created he him; male and female created he them.
28. And God blessed them, and God said unto them, Be fruitful, and multiply, and replenish the earth, and subdue it: and have dominion over the fish of the sea, and over the fowl of the air, and over every living thing that moveth upon the earth.
29. And God said, Behold, I have given you every herb bearing seed, which is upon the face of all the earth, and every tree, in the which is the fruit of a tree yielding seed; to you it shall be for meat.
30. And to every beast of the earth, and to every fowl of the air, and to every thing that creepeth upon the earth, wherein there is life, I have given every green herb for meat: and it was so.
31. And God saw every thing that he had made,

and, behold, it was very good. And the evening and the morning were the sixth day.

Millions of years lapsed between primitive primates and mankind. A primitive primate called a "necrolemer" was, in all probability, a common ancestor of man and ape. Some people may feel indignation regarding mankind's kinship with other earthly life, but God has revealed that the most common beginning of all natural life was dust. Indignation or not, God has also revealed that the most common ending of all natural life will be dust!

Hominids were the primates that very closely resembled man and they appeared somewhere between the "Cretaceous period, of the Mesozoic era," of about 120 million years ago; and the "Tertiary period, of the Cenozoic era," of about 30 million years ago. By the succession of increased intelligence, hominids evolved towards high specialization. They are divided into two groups: Group I: *Ramapithecines to Homo erectus*[9] and Group II: *Peking man to Cro-Magnon man.*[10] Even though the second group of hominids is reckoned by science as having been man; the fixed or highly specialized state of man is not reached until Cro-Magnon man appears. He is the only one listed in the second group from above that can justifiably be called "man." He was man in the finished, natural state. He was made in the "image" of God, but *not* in the "likeness" of God. Cro-Magnons were progenitors of Adam and preceded him in the last creative day, perhaps as far back as 50 thousand years ago, or their species may have been completed with the parents of Adam, six or seven thousand years ago. All the other hominids were creatures of transition, and were phased out as evolution progressed towards Cro-Magnon man. The evolutionary process of creating natural life ended after Cro-

Magnon man was made.

When God said, "Let us make man in our image, after our likeness," He was differentiating between the words *image*, meaning the natural body and *likeness*, meaning the spiritual body. Jesus was relating to both words when He said, "... he that hath seen me hath seen the Father ..." (John 14:9). On the Mount of Transfiguration, Peter, James, and John saw the spiritual bodies (not glorified, resurrected bodies) of Moses and Elias with Jesus (Matthew 17:1-13). What they saw confirms to us that spiritual bodies have some characteristics that resemble the natural bodies they once inhabited. One definite characteristic of Moses and Elias: they were men, not children, when they died. Therefore, the natural image of Jesus tells us that our Heavenly Father's Spirit has the image of a natural man, excluding, of course, spiritual light and other attributes of being divine.

God gave Cro-Magnon man dominion over every kind of earthly life, and it stands to reason since he was truly man, made in the "image" of God. The order of creation as such, bespeaks of Almighty God's ingenious way of providing food for His creatures, which caused an ecological balance in the earth. God was very pleased with what He had created: "For thou art great, and doest wondrous things: thou art God alone" (Psalms 86:10).

The sixth day of creation ended! It had been a very busy and long, creative working week and it was time to rest. But wait! God tells of His seventh day of rest from making everything; then He gives a summary of His creative work:

Chapter 2

1. Thus the heavens and the earth were finished, and all the host of them.
2. And on the seventh day God ended his work which he had made; and he rested on the seventh day from all his work which he had made.
3. And God blessed the seventh day, and sanctified it: because that in it he had rested from all his work which God created and made.

God created many different kinds of creatures before He finally reached His goal of creating man, and did so by the intricate process of evolution. Afterward, He blessed and sanctified the seventh day in which He rests from all His works of creation. "Sanctified" means set apart as being special. Today is a sanctified day, for we are still living in God's seventh day of rest, and will be until the end of time! God may still be working, but He is no longer doing a creative work. The evolutionary process of creating man in His "image" probably ended somewhere between 50 thousand and 6 thousand years ago. The process of making Eve from Adam was only a reproductive act. The process of making Adam into a "Son of God" by the breath of God, was a different kind of divine act; the act of making man in the "likeness" of God, which happened about six thousand years ago. Since all species are finished and fixed, and life can no longer evolve to a higher plane, then all organic life, in this day of rest, comes into the earth by reproducing "after his kind" or by "hybridization."

Biblical genealogies are records for our enlightenment and indicate that Adam was born about six thousand years ago. Archeologists are scientists who help enlighten truth. Many of them have found bones of prehistoric man and animals in the earth. To hold on to a traditional belief, an antagonist once said: "The devil put those bones there." It would be outside the nature of God to allow a spiritual entity, of the Kingdom of Darkness or Light to produce or change the natural remains left in the earth, just to destroy the credibility of true science, because true science and the word of God are compatible.

4. These are the generations of the heavens and of the earth when they were created, in the day that the Lord God made the earth and the heavens,
5. And every plant of the field before it was in the earth, and every herb of the field before it grew: for the Lord God had not caused it to rain upon the earth, and there was not a man to till the ground.
6. But there went up a mist from the earth, and watered the whole face of the ground.
7. And the Lord God formed man of the dust of the ground, and breathed into his nostrils

the breath of life; and man became a living soul.

The fourth through the seventh verses give a view of the preceding six creative days, which lasted for generations of time! If God's creative days lasted for that length of time, then it is quite evident that those days were not six literal days. God foreordained to create man, but if man had been created first, he would not have had anything to do. With no vegetation, there would have been no tilling. God prepared the earth and caused vegetation of the seas to spread upon land. God spoke during each creative day and caused life to successively rise to complexity, until man developed.

Cro-Magnons were "multiplying, replenishing, and subduing the earth," just as Genesis 1:28 says. They were created in the image (natural body) of God, created both male and female. Cro-Magnons were highly specialized and intelligent enough to make tools, paint, carve, and do many other things. However, without a spiritual body, they were not capable of worshipping God, committing sin, or speaking a human type language. They undoubtedly grieved their dead to some extent, and sometimes buried them with earthly goods, thinking they may revive; but that does not mean that they had a conscience towards God and belief in an afterlife.

Genesis 2:7 tells us that God formed man out of the dust of the ground, even though it took billions of years to do so. Since time seemed to be of no consequence to God as He created, then it is not unreasonable that about 50 thousand years may have passed between creating man in His image (natural body), and Adam's birth.

Adam was an offspring of Cro-Magnon man, because Adam was born a baby, just like all of mankind. God breathed into baby Adam's nostrils, and man became a "living soul" made in the "likeness" of God, therefore, a *Son of God*. Adam's spiritual body emanated from the breath of God, but his natural body was conceived in the uterus of a woman of the "soulless seed" of mankind. The essence of God's breath is great, powerful, and everlasting. When God breathed into Adam's nostrils, Adam received a spiritual body and the breath of God permeated the cells of Adam's reproductive mechanisms. This enabled Adam to transmit unto his descendants a spiritual body, which is by inheritance. Cro-Magnons did not have spiritual bodies, because they preceded Adam. Their nature was somewhat barbaric like other primates of the animal kingdom, because they were "flesh" only! They were at home in the earth, which is only a natural domain. Their feelings of loyalty, anger, grief, etc., were as other animals—instinctive and shallow, since they did not have spiritual brains or souls where deeper thoughts and feelings are manifested. If the brain or mind of a nonhuman animal is ever referred to as the soul of the animal, it must be regarded as a dead soul, and not a "living soul" (Genesis 2:7), like that which belongs to humans. That is why it is not considered murder to kill animals. However, when the life of a human being is taken, the earthly dwelling place of that person's spiritual body is destroyed, until the resurrection, for the spiritual body cannot die. The human, spiritual body develops along with the natural body, for both are inherited from Adam and Eve! The flesh and spirit must develop together, before being able to live in the earth. Therefore, it must be concluded that human embryos are only potential human beings, because they have not developed. Christ is the only Spirit that pre-existed His human body, because His divine Spirit was not inherited from Adam and Eve. Therefore, an earthly father did not beget Him. Consequently, it will be a gift from God if stem cells can be used to cure diseases and help eliminate some human suffering.

There are some fables that insinuate that earthly animals may dwell in heaven after they die. Impossible! Only a body with a "living soul" has the capacity to worship God. It is by our spiritual bodies that Christ will "quicken" our mortal bodies to resurrect or change, and become immortal: "Neither give heed to fables and endless genealogies, which minister questions, rather than godly edifying which is in faith: so do" (I Timothy 1:4).

The ability to speak a human language comes from the spiritual body! Adam was the first man with a spiritual body and God gave him the privilege of naming all the animals. Adam could speak, because he was made in the "likeness" of God who could speak. Some other species of animals evolved with physical attributes for speaking, such as a voice box, tongue, teeth, etc., yet none of them can speak or be taught to speak a human type language. This is not to be confused with certain birds that speak by mimicry, which is not with understanding. Animals may be able to communicate with humans and each other to a certain extent. Some of them can be taught to do some things by the method of rote, but never to speak a vocal language with understanding.

Some Biblical verses substantiate the fact that the ability to speak a human type language with understanding comes through the physical or natural mouth, from the spiritual body, which is called the "heart":

> O generation of vipers, how can ye, being evil, speak good things? For out of the abundance of the *heart* the mouth speaketh. A good man out of the good treasure of the *heart* bringeth forth good things: and an evil man out of the evil treasure bringeth forth

evil things. But I say unto you, That every idle word that men shall speak, they shall give account thereof in the day of judgment. For by thy words thou shalt be justified, and by thy words thou shalt be condemned (Matthew 12:34-37).

8. And the Lord God planted a garden eastward in Eden; and there he put the man whom he had formed.
9. And out of the ground made the Lord God to grow every tree that is pleasant to the sight, and good for food; the tree of life also in the midst of the garden, and the tree of knowledge of good and evil.
10. And a river went out of Eden to water the garden; and from thence it was parted, and became into four heads.
11. The name of the first is Pison: that is it which compasseth the whole land of Havilah, where there is gold;
12. And the gold of that land is good: there is bdellium and the onyx stone.
13. And the name of the second river is Gihon: the same is it that compasseth the whole land of Ethiopia.
14. And the name of the third river is Hiddekel: that is it which goeth toward the east of Assyria. And the fourth river is Euphrates.
15. And the Lord God took the man, and put him into the garden of Eden to dress it and to keep it.
16. And the Lord God commanded the man,

saying, Of every tree of the garden thou mayest freely eat:

17. But of the tree of the knowledge of good and evil, thou shalt not eat of it: for in the day that thou eatest thereof thou shalt surely die.

Adam became the first natural being with a spiritual body, and he needed a safe habitat that fitted his full nature, and that place was the Garden of Eden. The garden was specially planted for a very special creation—the "Sons of God." So God put Adam in the garden to protect him from the harshness of the earth outside. The Creator caused spiritual trees to grow out of the literal ground of the garden and the fruit of the trees were made for the Sons of God. That is why Genesis 1:29 related primarily to Cro-Magnon man. They were not Sons of God, and the literal food outside the Garden of Eden was meant for them. Adam and his descendants did not need earthly food that grew outside the garden when they had special food grown for them on the inside.

18. And the Lord God said, It is not good that the man should be alone; I will make him an help meet for him.
19. And out of the ground the Lord God formed every beast of the field, and every fowl of the air; and brought them unto Adam to see what he would call them: and whatsoever Adam called every living creature, that was the name thereof.
20. And Adam gave names to all cattle, and to the fowl of the air, and to every beast of the

21. field; but for Adam there was not found a help mate for him.
21. And the Lord God caused a deep sleep to fall upon Adam, and he slept: and he took one of his ribs, and closed up the flesh instead thereof;
22. And the rib, which the Lord God had taken from man, made he a woman, and brought her unto the man.
23. And Adam said, This is now bone of my bones, and flesh of my flesh: she shall be called Woman, because she was taken out of man.
24. Therefore shall a man leave his father and his mother, and shall cleave unto his wife: and they shall be one flesh.
25. And they were both naked, the man and his wife, and were not ashamed.

In this passage the writer reiterated that God created all life from the dust of the ground. Animals, including Cro-Magnon man could not see the spiritual side of the Garden of Eden, because they did not have spiritual eyes. God said that man should not be alone, and of all the creatures in the animal kingdom, there was not one that was right for Adam. God planned to conjugally join Adam with a "helpmate" and she would have to be made in the "likeness" of God, as was Adam. She would have to possess a spiritual body! God caused Adam to fall into a deep sleep. God then opened Adam's side and caused one of his ribs to grow a uterus in which Eve's body developed. God closed Adam's side, woke him, and presented baby Eve to him!

Each act of God's creative will was but once! This explains why Eve was made from Adam! God had already breathed into Adam's nostrils the breath of life that gave him a "living soul," and God performed this work only one time! Eve inherited a natural and spiritual body from Adam! Therefore, she was made into a "helpmate" for Adam; able to help reproduce *Sons of God.* Adam said Eve was: "bone of his bones, and flesh of his flesh," and she truly was in the literal sense. Adam also said she would be called "woman," because she was taken out of the "womb" (uterus) of man.

Very far back when life began, reproduction occurred asexually by fission. Then when sexual reproduction occurred, it could be clearly seen in many species that males and females have similar characteristics that remain underdeveloped. This indicates that when God started sexual reproduction, He first had to separate the one-parent organisms into two parts: males and females. When the two are separate, each one is only one-half of a whole, because reproduction cannot occur. When God ordained the married state in the Garden of Eden, He was pronouncing a blessed event that fuses the two halves together into one flesh. The "one flesh" (a baby) actually occurs when the reproductive germ cells of each half are united and developed.

Our first parents lived in a state that no one else has ever lived in. They did not have fallen parents preceding them, as we have of them. Therefore, Adam and Eve were formed without a fallen nature, and because of their innocence, they were unaware of their nakedness. After God brought mankind into the earth, the fullness of creation was made manifest, so the six cosmic days of creation ended:

> He hath made every thing beautiful in his time: also he hath set the world in their heart, so that no

man can find out the work that God maketh from the beginning to the end (Ecclesiastes 3:11).

Chapter 3

1. Now the serpent was more subtle than any beast of the field which the Lord God had made. And he said unto the woman, Yea, hath God said, Ye shall not eat of every tree of the garden?
2. And the woman said unto the serpent, We may eat of the fruit of the trees of the garden:
3. But of the fruit of the tree which is in the midst of the garden, God hath said, Ye shall not eat of it, neither shall ye touch it, lest ye die.
4. And the serpent said unto the woman, Ye shall not surely die:
5. For God doth know that in the day ye eat thereof, then your eyes shall be opened, and ye shall be as gods, knowing good and evil.
6. And when the woman saw that the tree was good for food, and that it was pleasant to the eyes, and a tree to be desired to make one wise, she took of the fruit thereof, and

did eat, and gave also unto her husband with her; and he did eat.

7. And the eyes of them both were opened, and they knew that they were naked; and they sewed fig leaves together, and made themselves aprons.

8. And they heard the voice of the Lord God walking in the garden in the cool of the day: and Adam and his wife hid themselves from the presence of the Lord God amongst the trees of the garden.

9. And the Lord God called unto Adam, and said unto him, Where art thou?

10. And he said, I heard a voice in the garden, and I was afraid, because I was naked; and I hid myself.

11. And he said, Who told thee that thou wast naked? Hast thou eaten of the tree, whereof I commanded thee that thou shouldest not eat?

12. And the man said, The woman whom thou gavest to be with me, she gave me of the tree, and I did eat.

13. And the Lord God said unto the woman, What is this that thou hast done? And the woman said, The serpent beguiled me, and I did eat.

14. And the Lord God said unto the serpent, Because thou hast done this, thou art cursed above all cattle, and above every beast of the field; upon thy belly shalt thou go, and dust shalt thou eat all the days of thy life:

15. And I will put enmity between thee and the woman, and between thy seed and her seed; it shall bruise thy head, and thou shalt bruise his heel.
16. Unto the woman he said, I will greatly multiply thy sorrow and thy conception; in sorrow thou shalt bring forth children; and thy desire shall be to thy husband, and he shall rule over thee.
17. And unto Adam he said, Because thou hast hearkened unto the voice of thy wife, and hast eaten of the tree, of which I commanded thee, saying, Thou shalt not eat of it: cursed is the ground for thy sake; in sorrow shalt thou eat of it all the days of thy life;
18. Thorns also and thistles shall it bring forth to thee; and thou shalt eat the herb of the field;
19. In the sweat of thy face shalt thou eat bread, till thou return unto the ground; for out of it wast thou taken: for dust thou art, and unto dust shalt thou return.
20. And Adam called his wife's name Eve; because she was the mother of all living.
21. Unto Adam also and to his wife did the Lord God make coats of skins, and clothed them.
22. And the Lord God said, Behold, the man is become as one of us, to know good and evil: and now, lest he put forth his hand, and take also of the tree of life, and eat, and live for ever:

23. Therefore the Lord God sent him forth from the garden of Eden, to till the ground from whence he was taken.
24. So he drove out the man; and he placed at the east of the garden of Eden Cherubims, and a flaming sword which turned every way, to keep the way of the tree of life.

The person of Satan appeared in the Garden of Eden disguised as a serpent. He cunningly reasoned with Eve by questioning her about the fruit trees. She answered him by saying that God had indeed said that they could eat the fruit of the trees, but signified that they should not eat of the fruit of the tree in the middle of the garden. She told Satan they would die, even if they touched it. Satan then told her that she would not die, but would become as gods, knowing good and evil. Satan tricked Eve by lying to her, "… for he is a liar, and the father of it" (John 8:44). He frequently succeeds in deceiving people by offering them a portion of the truth! He tricked Eve and she reacted contrary to the will of God. She ate of the forbidden fruit and then offered it to her husband, who also ate. God loved Adam and Eve and did not will for them to die. If they had not eaten of the forbidden fruit, then their spiritual and natural bodies would have lived forever in the Kingdom of Light, for they would have eventually eaten of the "Tree of Life." God made the flesh of Adam and Eve from the dust of the earth like the rest of natural life. Consequently, their flesh was corruptible—capable of dying when they were created, but if they ate of the forbidden fruit, they would "surely" die in a physical way and fall in a spiritual way (Genesis 2:17). They had opportunity to eat of the "Tree of Life" first, if only they had done so.

When Satan was able to get Eve to look upon the fruit, it appealed to her carnal nature, so that Satan's lie, coupled with the lust of the flesh, caused a strong temptation that was hard to resist:

> For all that is in the world, the lust of the flesh, and the lust of the eyes, and the pride of life, is not of the Father, but is of the world. And the world passeth away, and the lust thereof: but he that doeth the will of God abideth for ever (I John 2:16,17).

A portion of truth that Satan revealed is that people do become as gods when their eyes are opened. They could see their nakedness because they had lost their innocence. They also had the ability to know and to choose to do good or evil. However, Adam and Eve did not realize that they were in the Dark Kingdom. That realization could come only by the law: "For I was alive without the law once: but when the commandment came, sin revived, and I died" (Romans 7:9). The same condition occurs every time a child grows and learns good and evil, which makes him accountable and no longer innocent. However, thank God, the law came and taught us, and the blood came and bought us, whenever we become born again (Galatians 3:24) and (I Corinthians 6:20)!

Satan tempted innocent Eve. She ate of the forbidden, spiritual fruit. She shared it with innocent Adam. This caused them both to fall into the Dark Kingdom. As a result, all of mankind genetically inherits the fallen nature from our first parents. God set down the standard of sin for He knew what would be harmful to the Sons of God. You may tell an innocent child not to accept candy from a stranger, but a sinister person could entice a young child to eat poisonous candy. The candy would be the direct cause of death. Neither Eve nor Adam knew that she had committed a sinful act, until after the act was done! Before the act, they were innocent. After the act, they were as gods, knowing good and evil, but not realizing that disobeying God is "exceeding sinful," for the realization could come only by the "Commandments" (law) (Romans 7:12,13).

Under the Dispensation of Grace, accountable people everywhere must be born-again, out of the Dark Kingdom and into the Kingdom of Light, by accepting Jesus Christ as Savior and repenting under His sacrificial, shed blood. Under the Dispensation of the Law, only accountable Israelites could repent under the blood of animal sacrifices, even though the blood did not perfectly purge them. Only their flesh was purified, for year after year they were still conscious of their sins:

> For if the blood of bulls and of goats, and the ashes of an heifer sprinkling the unclean, sanctifieth to the purifying of the flesh: How much more shall the blood of Christ, who through the eternal Spirit offered himself without spot to God, purge your conscience from dead works to serve the living God (Hebrews 9:13,14)?

The Lord was walking in the Garden of Eden at twilight time when He called out to Adam. God knew that Adam and his wife were hiding, and He was gently prodding them out. Adam truthfully told the Lord God that he was afraid because he was naked. Adam had that realization for the first time, for eating of the forbidden fruit gave him awareness. However, at the time, Adam apparently did not know the kind of fruit he was eating, for Scripture says: "And Adam was not deceived, but the woman being deceived was in the transgression" (I Timothy 2:14). Eating of the forbidden fruit made them both fall into the Dark Kingdom, but Eve, not Adam, committed the sin. However, before the law there was no "place of repentance," so sin was not imputed: "For until the law sin was in the world: but sin is not imputed when there is no law" (Romans 5:13). God did not impute sin before the law, but He did pronounce judgment upon Satan and mankind for what happened.

Eve frankly told the Lord God that she ate of the forbidden fruit because Satan deceived her. God then placed a curse upon Satan. Figuratively speaking, He placed him lower than any animal—to crawl in and eat dust all his life, which is forever. When the universe turns back into dust, it will end up in the realm of the cursed: "... and dust shall be the serpent's meat..." (Isaiah 65:25). When God judged Satan, He gave him the materialistic realm (ultimately dust) for an inheritance: "Of judgment, because the prince of this world is judged" (John 16:11).

God declared warfare between Satan and his seed and Eve and her seed when He said: "... it shall bruise thy head, and thou shalt bruise his heel" (Genesis 3:15). When Adam and Eve ate of the forbidden fruit, Satan was victorious in his pursuit. After Adam and Eve lost their innocence they could do good or evil in the sight of God. God planned a way to redeem them and their seed, but they would have to battle. If the woman and her seed choose to do that which is good in the sight of God, Satan receives a "bruised head." However, Satan has a strong tool called "temptation" which can cause the woman and her seed to choose evil. The temptation is analogous to a "bruised heel," which can cause the woman and her seed to fall to the temptation if it is not endured and overcome.

Jesus was born human and divine, and is a God knowing

good and evil. Since God gave Satan permission to tempt humankind in Genesis 3:15, then Satan had permission to also tempt Jesus! It caused Jesus to be able to identify with the rest of humankind by being able to feel the power of temptation but having divine power to resist, since He did not have a fallen nature because His Spiritual body pre-existed Adam. Concerning people with accountability, the Bible shows that with the exclusion of Jesus, "… all have sinned, and come short of the glory of God" (Romans 3:23). It is the nature of a person in the Dark Kingdom to sin. That is why a person must enter the Light Kingdom by being born-again! Then they can have the power of God to help them resist temptation.

The women of the earth inherited a penalty because of Eve's sin. From that time and on down through the ages, women would suffer in bearing their children; the desire for their husbands being irrelevant to their suffering. Humanly and genetically speaking, Sons of God males rule over females, because males carry in their bodies the factor that determines the sex of their children. Males also carry in their bodies the genetic factor that generates a spirit body, but Sons of God females are genetically needed helpers.

God pronounced other penalties that would occur to humankind because of the fall. Man would have to struggle with the earth's wild growth for the ground to produce food, and the struggle would be hard enough to cause perspiration. He would also have to experience the death of the natural body, which was already well known outside the Garden of Eden.

Adam gave his helpmate the name "Eve;" a name that means that she is the "mother of all living" (Genesis 3:20). The connotation relates to the inherited nature of possessing a spiritual body that has a soul, and was made to distinguish between the descendants of Eve, and the soulless seed of mankind and other earthly creatures that existed at that time outside the garden. The soulless seed of mankind were like the beasts of the fields that naturally live and breathe, but Eve was not their mother.

The bodies of our first parents were partially covered with fig leaves, but leaves were not sufficient for warmth outside the garden, so God provided animal skins for them to wear. God did not have to kill animals for their skins! The cycle of being born and dying had been consistent in the earth since God started life billions of years ago, so there were plenty of dead animals whose skins were available for clothing. Some preachers and teachers of the gospel have erroneously proclaimed that God killed the animals for their skins, believing it was a foresight of blood sacrifices. God has meant for death to occur to all life in its natural state. There are some people who may not have enough compassion for others, but have too much compassion for animals—to a fault. They have organized to protect animals (which is not bad in itself), but whenever they protest the use of animals when it is for the benefit of mankind, they are stepping into a territory where God has already shown us our privileges concerning them. First of all, God gave man dominion over the animal kingdom. The assurance we already have to use them for food, and also their skins and fur for warmth, like He did for Adam and Eve should not be protested. These privileges show us that God sanctions any way mankind can benefit from the use of animals. Unnecessary, brutal, and unlawful treatment of animals is sinful and does not benefit mankind. Animals do not have "living souls," and common sense shows us that humans should know right from wrong concerning them. There is no other sure foundation on which to stand, other than the word of

God, when discerning right from wrong.

In this passage the pronoun "him" (Genesis 3:23), refers to both Adam and Eve, as God regarded them as one. He sent them out of the garden because He could not trust them to leave the "Tree of Life" alone. Eating its fruit would have caused them to live forever in the fallen state, like the fallen angels, and God would not have had opportunity to redeem them. Satan would have been an eternal victor! Their souls would have been bound in the realm of spiritual darkness and their natural bodies would have been bound in the state of dust forever.

God is a merciful and loving Father, and we can thank and praise Him for redeeming men from Satan whenever men accept the death and resurrection of the Lord, Jesus Christ. God drove out our first parents from the garden, to the earth from which Adam was taken, to till the ground for sustenance—never again in this life to eat the very special food that was created for them. The Garden of Eden is a dual dimension, created for the dual bodied man. It still exists in the Middle East, but "Cherubim and a flaming sword" are the spiritual forces that block the vision of all, "… to keep the way of the tree of life" (Genesis 3:24).

Chapter 4

1. And Adam knew Eve his wife; and she conceived, and bare Cain, and said, I have gotten a man from the Lord.
2. And she again bare his brother Abel. And Abel was a keeper of sheep, but Cain was a tiller of the ground.
3. And in process of time it came to pass, that Cain brought of the fruit of the ground an offering unto the Lord.
4. And Abel, he also brought of the firstlings of his flock and of the fat thereof. And the Lord had respect unto Abel and to his offering:
5. But unto Cain and to his offering he had not respect. And Cain was very wroth, and his countenance fell.
6. And the Lord said unto Cain, Why art thou wroth? and why is thy countenance fallen?
7. If thou doest well, shalt thou not be accepted? and if thou doest not well, sin lieth at the door. And unto thee shall be

> 　　　　his desire, and thou shalt rule over him.
> 8.　　And Cain talked with Abel his brother: and it came to pass, when they were in the field, that Cain rose up against Abel his brother, and slew him.

The remaining years of Adam and Eve began outside the Garden of Eden in a place contrary to their full nature. The Lord blessed Eve, and she conceived and gave birth to their first-born son, and they called him "Cain." She gave birth to another son, and they called him "Abel." Since the Bible does not state that Eve conceived a second time, it is possible that Cain and Abel were twins. The brothers grew and became endowed in honorable jobs: Abel a shepherd, and Cain a gardener.

Our first family realized the love and tender mercies of God. In appreciation, they offered first fruits of their labor to Him. God had respect for Abel's offering, but He did not have respect for Cain's offering, because Cain had not done his best! It has been proclaimed that Cain's offering was rejected because it was not a blood offering. Verse seven plainly states that if Cain had committed himself to doing his best, he also would have been accepted by God and would not have endured dissatisfaction, which opened the door to jealousy. The jealously drove him to kill his brother.

> 9.　　And the Lord said unto Cain, Where is Abel thy brother? And he said, I know not: Am I my brother's keeper?
> 10.　 And he said, What hast thou done? the voice of thy brother's blood crieth unto me from the ground.
> 11.　 And now art thou cursed from the earth,

	which hath opened her mouth to receive thy brother's blood from thy hand;
12.	When thou tillest the ground, it shall not henceforth yield unto thee her strength; a fugitive and a vagabond shalt thou be in the earth.
13.	And Cain said unto the Lord, My punishment is greater that I can bear.
14.	Behold, thou hast driven me out this day from the face of the earth; and from thy face shall I be hid; and I shall be a fugitive and a vagabond in the earth; and it shall come to pass, that every one that findeth me shall slay me.
15.	And the Lord said unto him, Therefore whosoever slayeth Cain, vengeance shall be taken on him sevenfold. And the Lord set a mark upon Cain, lest any finding him should kill him.

God was not surprised by the sin of murder, as anyone who has a fallen nature is capable of committing any sin. God knew that Abel's natural body was dead, for He heard the voice of his spirit cry to Him. The Lord God questioned Cain of the whereabouts of his brother. Cain lied and then tried to rationalize by asking the Lord if he was his brother's keeper. Cain knew that he had chosen to do evil by killing Abel, but he did not know that evil was dark and deadly, and that spiritually, he was in the Dark Kingdom. His sin was not imputed to him because the law had not yet been given: "… for where no law is, there is no transgression" (Romans 4:15). However, that does not mean that he did not commit a transgressing act, only that the transgression at that time was considered nil. Therefore, if sin was not imputed before the law, then before the law there was no "place of repentance."

God allowed a curse of drought upon the earth. Although Cain was experienced in tilling, his efforts to produce food would be difficult. He would have to wander to find enough food to sustain himself. Having to leave the place where his family dwelled and being hidden from the face of God caused Cain to despair. The soulless seed of mankind were barbaric, and he feared his life would end at their hands. God put a mark upon Cain that would help protect him, but it did not mean that he was exempt from being killed. However, his death would be avenged "sevenfold" if anyone did kill him. The soulless seed of mankind were intelligent enough to fear the mark, but like other animals, they were sometimes unpredictable. They were not conscientious about loving or killing, for they did not know good and evil.

16. And Cain went out from the presence of the Lord, and dwelt in the land of Nod, on the east of Eden.
17. And Cain knew his wife; and she conceived, and bare Enoch: and he builded a city, and called the name of the city, after the name of his son, Enoch.
18. And unto Enoch was born Irad: and Irad begat Mehujael: and Mehujael begat Methusael: and Methusael begat Lamech.

Cain left the dwelling place of his parents and went into the land of Nod, located in the east of Eden. There he met his wife! She was a woman of the "soulless seed" of mankind, a Cro-

Magnon woman and not a descendant of Adam and Eve! Eve being the "mother of all living" refers to her soul filled descendants. All earthly lives that preceded Adam and Eve were mortal. Without souls, men were mortal like other animals, but still men. Adam and Eve were mortal, so their flesh was already capable of dying, before the fall. The natural side of man being mammal expresses a wondrous order of having created him as the most advanced mammal in the last day of creation.

Cain's wife conceived and gave birth to a son and they called him "Enoch." Cain built a city and named it the same, in honor of his son. Enoch was a "corrupt seed"—the essence of a life that was a "degradation" to God's creation (Genesis 6:12), because his parents were very different: His father had a soul, but his mother was soulless! She had not been a "helpmate" to Cain, for she could not help him build a spiritual body. Their union brought a life into the earth that corrupted God's creative work.

In the Biblical passages below, the genealogy of Cain is given down to his fifth generation, and all his descendants were "corrupt seed." They had some speech ability because they were partly descended from Cain who had a spiritual body.

> 19. And Lamech took unto him two wives: the name of the one was Adah, and the name of the other Zillah.
> 20. And Adah bare Jabal: he was the father of such as dwell in tents, and of such as have cattle.
> 21. And his brother's name was Jubal: he was the father of all such as handle the harp and organ.

22. And Zillah, she also bare Tubalcain, an instructor of every artificer in brass and iron: and the sister of Tubalcain was Naamah.
23. And Lamech said unto his wives, Adah and Zillah, Hear my voice; ye wives of Lamech, hearken unto my speech: for I have slain a man to my wounding, and a young man to my hurt.
24. If Cain shall be avenged sevenfold, truly Lamech seventy and sevenfold.

Cain's descendants grouped together in family units. Cain must have brought the idea of marriage into the land of Nod, even though most species naturally group together with their own kind. Whatever kind of nature the corrupt seed had, they were not in totality just an animal because they partly descended from Cain, who had a spiritual body. Lamech, who was Cain's fifth generation descendant, had two wives who bore him two children apiece: Adah bore Jabal—the first tent dweller, and Jubal—the first musician of the harp and organ. Zillah bore Tubalcain—a teacher in the art of making iron and brass products, and a daughter named Naamah.

Cain was a special person in the land of Nod because of the mark that God placed upon him. Cain undoubtedly explained the mark to his descendants, and they undoubtedly revered him to some extent. However, it was not an indication that he would not be struck down by a crude and quick-tempered man of the soulless seed or a corrupt descendent of one.

Lamech returned home to his wives one day and announced to them that he had become wounded when he killed a man. He stressed to them that the man was young, and the consequence of killing him would be a hurt worse than his wound. Lamech had killed Cain! Out of dread and dull reasoning, Lamech tried to justify himself by believing that killing a special person had made him "special" too. He surmised that if Cain's death would be avenged "sevenfold," then truly Lamech's death by someone's hand would be avenged "seventy and sevenfold." When God said that Cain's death would be avenged sevenfold if anyone killed him, then it would truly come to pass, but the words that Lamech spoke would not come to pass! Adam, Eve, and Cain were the only Sons of God who were alive on the earth at that time since Adam's third son, Seth, had not yet been born. Therefore, Cain had to be the one whom Lamech killed. The Sons of God were the lives that were of the most significance to God, for they were the only lives on earth that had spiritual bodies. Cain was considered a young man when he was slain, as men in those days lived to be hundreds of years old. Cain may have had descendants other than those listed in the Bible, but those listed down to his sixth generation were the only ones who were born during his lifetime.

> 25. And Adam knew his wife again; and she bare a son, and called his name Seth: For God, said she, hath appointed me another seed instead of Abel, whom Cain slew.
> 26. And to Seth, to him also there was born a son; and he called his name Enos: then began men to call upon the name of the Lord.

Cain and Abel were dead! Adam and Eve needed another offspring so that their kind of seed would remain. Eve conceived and gave birth to her third child, and they called him

Seth. When Eve said that Seth was to take the place of Abel, she was verifying that a son like Abel would be an upright son who God would respect. The word of God states that our first parents had other sons and daughters after Seth was born (Genesis 5:4), but Seth would be the line through which Adam and Eve's seed would remain.

Seth was one hundred and five years old before he had an offspring named Enos. Without a doubt, Seth's wife had to be a relative, an offspring from one of his brothers or sisters who were born after him, because the bloodline from Adam and Eve through Seth to Noah was untainted. The bloodline through Cain became tainted by mixing with the soulless seed of mankind.

"… Then began men to call upon the name of the Lord" (Genesis 4:26). Of this saying, Adam Clarke meant that these men began to call themselves Sons of God, but it did not mean that men began to pray to God:

> "Then began men to call themselves by the name of the Lord" which words are supposed to signify that in the time of Enos the true followers of God began to distinguish themselves, and to be distinguished by others, by the appellation of sons of God.[11]

Chapter 5

1. This is the book of the generations of Adam. In the day that God created man, in the likeness of God made he him;
2. Male and female created he them; and blessed them, and called their name Adam, in the day when they were created.

The genealogy of Adam was written for our understanding. It begins with Adam, a man born with only a natural body, who afterwards, by the breath of God became a living soul, a Son of God. This is when God made man in his "likeness," which means the spiritual body. God is a Spirit, but we would not know what God's spiritual body looks like unless we could compare it with the image of something else. Jesus said: "… he that hath seen me hath seen the Father …" (John 14:9). Therefore, we do not have to think of God as having wings, fins, horns, paws, etc., because we know His Spirit has the image of a man. Moses was privileged to see the backside of God: "And I will take away mine hand, and thou shalt see my back parts: but my face shall not be seen" (Exodus 33:23). This verse substantiates the fact that God does have a back, face, and hands, just like mankind made in His image.

God blessed our first parents and named them Adam, (which means mankind). They are the parents of all living souls, (which are spirit-filled bodies), made in His image and likeness.

> 3. And Adam lived an hundred and thirty years, and begat a son in his own likeness, after his image; and called his name Seth:
> 4. And the days of Adam after he had begotten Seth were eight hundred years: and he begat sons and daughters:
> 5. And all the days that Adam lived were nine hundred and thirty years: and he died.
> 6. And Seth lived an hundred and five years, and begat Enos:
> 7. And Seth lived after he begat Enos eight hundred and seven years, and begat sons and daughters:
> 8. And all the days of Seth were nine hundred and twelve years: and he died.

Adam was one hundred and thirty years old when Seth, his third child was born. Seth was a Son of God, like Adam, possessing both a natural and spiritual body. Although Adam and Eve had other children, only their first three sons were named in their genealogy. Therefore, we can see by the written genealogy that the human family was passed down through Seth, the third child and son of Adam and Eve.

When God breathed into Adam's nostrils, Adam received a spiritual body that cannot die. However, Adam's natural body was mortal and the day appeared when he took his last natural breath and died. Many people in Adam's day lived very long lives and begat many offspring. Adam and Eve had other children after Seth was born. Likewise, Seth had other children after Enos was born; then Seth grew old and died.

The following verses of genealogies give a straight bloodline descent from Enos, the son of Seth, to Noah and his sons. The order and consistency of those genealogies can be seen as they give the names of the first-born sons. They then mention that other sons and daughters were born after them. This same consistency can be seen in Genesis 5:4, as it relates to Adam and Eve having other sons and daughters born, but only after the birth of Seth. Those written genealogies also verify the fact that Cain's wife had to be a Cro-Magnon woman—a woman of the soulless seed of mankind.

9. And Enos lived ninety years, and begat Cainan:
10. And Enos lived after he begat Cainan eight hundred and fifteen years, and begat sons and daughters:
11. And all the days of Enos were nine hundred and five years: and he died.
12. And Cainan lived seventy years, and beget Mahalaleel:
13. And Cainan lived after he begat Mahalaleel eight hundred and forty years, and begat sons and daughters:
14. And all the days of Cainan were nine hundred and ten years: and he died.
15. And Mahalaleel lived sixty and five years, and begat Jared:
16. And Mahalaleel lived after he begat Jared

eight hundred and thirty years, and begat sons and daughters:
17. And all the days of Mahalaleel were eight hundred ninety and five years: and he died.
18. And Jared lived an hundred sixty and two years, and he begat Enoch:
19. And Jared lived after he begat Enoch eight hundred years, and begat sons and daughters:
20. And all the days of Jared were nine hundred sixty and two years: and he died.
21. And Enoch lived sixty and five years, and begat Methuselah:
22. And Enoch walked with God after he begat Methuselah three hundred years, and begat sons and daughters:
23. And all the days of Enoch were three hundred sixty and five years:
24. And Enoch walked with God: and he was not; for God took him.

Enoch was the father of Methuselah. Enoch was three hundred and sixty-five years old when God translated him into another realm:

By faith Enoch was translated that he should not see death; and was not found, because God had translated him: for before his translation he had this testimony, that he pleased God (Hebrew 11:5).

God translated Enoch by placing him in another realm, but that place was not in Heaven where God dwells. Enoch inherited a fallen nature from Adam and Eve like all men.

Therefore, his spirit was not purged, and neither could his flesh enter into God's realm:

> Now this I say, brethren, that flesh and blood cannot inherit the kingdom of God: neither doth corruption inherit incorruption (I Corinthians 15:50).

Enoch's translation happened in the same way that it will happen to all who are alive when Christ returns: to change from a corruptible body into an incorruptible body one would literally die, but the change would be so rapid that one would miss the throes of death, for Scripture says: "... it is appointed unto men once to die, but after this the judgment" (Hebrews 9:27). Therefore, Enoch died, but when you are changed to dust in the twinkling of an eye, you do not *see* death. Enoch's spirit is at rest in Abraham's Bosom. His corruptible flesh went back to dust and went into the ground from whence it came. He was changed to dust, but he did not receive a glorified body because his change happened before Jesus' death and resurrection:

> And he is the head of the body, the church: who is the beginning, *the firstborn of the dead;* that in all things he might have the pre-eminence (Colossians 1:18).

> 25. And Methuselah lived an hundred eighty and seven years, and begat Lamech.
> 26. And Methuselah lived after he begat Lamech seven hundred eighty and two years, and begat sons and daughters:
> 27. And all the days of Methuselah were nine hundred sixty and nine years: and he died.

28. And Lamech lived an hundred eighty and two years, and begat a son:
29. And he called his name Noah, saying, This same shall comfort us concerning our work and toil of our hands, because of the ground which the Lord hath cursed.
30. And Lamech lived after he begat Noah five hundred ninety and five years, and begat sons and daughters:
31. And all the days of Lamech were seven hundred seventy and seven years: and he died.
32. And Noah was five hundred years old: and Noah begat Shem, Ham and Japheth.

Enoch named his son "Methuselah" which means: "… at my death the waters will come …."[12] Methuselah probably lived longer than anyone else on earth. Since he was untainted in his bloodline, the flood of God's wrath did not take him out of the earth. He died in some natural way on or near the day the great flood began. Lamech died five years before his father Methuselah, so Noah and his family were the only justified people left in the earth that could enter into the ark.

Noah obeyed God! Noah was used by God to bring about His will upon the earth. The name Noah was a comfort to the people. God allowed a curse of drought upon the ground, and because of Noah's obedience, God was able to send the great flood that restored it.

Noah was five hundred years old when his three sons, Shem, Ham, and Japheth, were born, indicating they were born triplets. They married women from the soul seed of mankind. Those women were helpmates, for they helped reproduce Sons of God.

Chapter 6

1. And it came to pass, when men began to multiply on the face of the earth, and daughters were born unto them,
2. That the sons of God saw the daughters of men that they were fair; and they took them wives of all which they chose.
3. And the Lord said, My spirit shall not always strive with man, for that he also is flesh: yet his days shall be an hundred and twenty years.
4. There were giants in the earth in those days; and also after that, when the sons of God came in unto the daughters of men, and they bare children to them, the same became mighty men which were of old, men of renown.

Cro-Magnons, the soulless seed of mankind, were multiplying on earth, and daughters were born to them. The Sons of God who were of the soul seed of mankind were attracted to them and took them for wives. Their offspring were corrupt seed, and their sons were famous for their size and strength. Adam Clarke said that the original word translated from the Septuagint literally describes those offspring as *"earth-born."*[13] In other words, their mothers were not helpmates to the Sons of God. Cro-Magnon women were not mothers of "all living" souls as was Eve but were earth mothers, made only of flesh.

The Bible does not mention the Daughters of God marrying the sons of men, because it was inconsequential. If they cohabited, their union produced only soulless seed—not corrupt seed. Daughters of God could not be helpmates to males who could not produce offspring with spiritual bodies.

The soulless seed of mankind could not sin because they did not have spiritual bodies, but they were a constant temptation to the Sons of God. God was tired of striving with the Sons of God because they were also "flesh" through which lust manifests itself. That is why He had to rid the earth of the flesh that tempted them. Noah had preached righteousness to the Sons of God, but they would not obey. God said that in a hundred and twenty years hence, He would never again have to deal with that particular sin because He would wipe out the cause of the temptation and the corruption that falling to the temptation brought about upon the earth.

> 5. And God saw that the wickedness of man was great in the earth, and that every imagination of the thoughts of his heart was only evil continually.
> 6. And it repented the Lord that he had made man on the earth, and it grieved him at his heart.

7. And the Lord said, I will destroy man whom I have created from the face of the earth; both man, and beast, and the creeping thing, and the fowls of the air; for it repenteth me that I have made them.
8. But Noah found grace in the eyes of the Lord.

God was displeased when He saw the Sons of God mix with the soulless seed of mankind. He repented that He had created "flesh," because "... the flesh profiteth nothing ..." (John 6:63). He did not repent for having made the "spirit," for the spirit is of the essence of God and cannot die. So when God repented—He changed His mind about allowing a certain kind of life to remain in the earth. He sent a worldwide deluge to wipe out Cro-Magnon man that was flesh only, the corrupt seed, and the defiled Sons of God from the earth. All other flesh that was of the dry land would also have to die in the flood, but a remnant of them would be preserved in the ark. Once again, the world would be free of any corruption to God's creation.

9. These are the generations of Noah: Noah was a just man and perfect in his generations, and Noah walked with God.
10. And Noah begat three sons, Shem, Ham, and Japheth.
11. The earth also was corrupt before God, and the earth was filled with violence.
12. And God looked upon the earth, and, behold, it was corrupt; for all flesh had corrupted his way upon the earth.
13. And God said unto Noah, The end of all

flesh is come before me; for the earth is filled with violence through them; and, behold, I will destroy them with the earth.

The Sons of God were preserved for the replenishment of the earth, as can be seen in Noah's genealogy. "Noah was a just man and perfect in his generations," because his bloodline had not been contaminated by the soulless seed of mankind: "... the root meaning for the Hebrew word for perfect is *whole or complete* ...."[14] Consequently, Noah's generations were "whole or complete" in their spiritual makeup. God could tolerate each seed of mankind in their place, but He could not tolerate them together. By mixing, they had corrupted "his way" on the earth, which was His creation, by producing a corrupt seed that He had not willed to exist. Therefore, they caused "violence" in the earth by desecrating the sacredness of God's perfect, creative work.

14. Make thee an ark of gopher wood; rooms shalt thou make in the ark, and shalt pitch it within and without with pitch.
15. And this is the fashion which thou shalt make it of: The length of the ark shall be three hundred cubits, the breadth of it fifty cubits, and the height of it thirty cubits.
16. A window shalt thou make to the ark, and in a cubit shalt thou finish it above; and the door of the ark shalt thou set in the side thereof; with lower, second, and third stories shalt thou make it.
17. And, behold, I, even I, do bring a flood of waters upon the earth, to destroy all flesh,

18. But with thee will I establish my covenant; and thou shalt come into the ark, thou, and thy sons, and thy wife, and thy sons' wives with thee.
19. And of every living thing of all flesh, two of every sort shalt thou bring into the ark, to keep them alive with thee; they shall be male and female.
20. Of fowls after their kind, and of cattle after their kind, of every creeping thing of the earth after his kind, two of every sort shall come unto thee, to keep them alive.
21. And take thou unto thee of all food that is eaten, and thou shalt gather it to thee; and it shall be for food for thee, and for them.
22. Thus did Noah; according to all that God commanded him, so did he.

God gave Noah precise instructions on how to build an ark. He made it three stories high with one window and one door. He made it waterproof with pitch. The measurements of the ark were given in cubits:

> In the Old Testament, lineal measure was based on the common cubit of 17.5 inches. Three kinds of cubits are known from ancient times: (1) In ancient Egypt a long cubit of 20.65 inches and a short cubit of 17.6 inches were used. (2) In Mesopotamia the "royal" cubit was 19.8 inches. (3) In the Old

Testament, besides the common cubit, Ezekiel mentions a long cubit of seven handbreadths of 20.44 inches (Ezek. 40:5, 42). In ancient Greece and Rome still other cubit measures were used.[15]

God told Noah that He would bring "a flood of waters upon the earth." He would also "establish a covenant" with Noah and allow him and his family to enter into the ark. It was Noah's responsibility to gather enough food to feed every life that was preserved for posterity. Noah was considered righteous because he obeyed God.

Chapter 7

1. And the Lord said unto Noah, Come thou and all thy house into the ark; for thee have I seen righteous before me in this generation.
2. Of every clean beast thou shalt take to thee by sevens, the male and his female: and of beasts that are not clean by two, the male and his female.
3. Of fowls also of the air by sevens, the male and the female; to keep seed alive upon the face of all the earth.
4. For yet seven days, and I will cause it to rain upon the earth forty days and forty nights; and every living substance that I have made will I destroy from off the face of the earth.
5. And Noah did according unto all that the Lord commanded him.
6. And Noah was six hundred years old when the flood of waters was upon the earth.

God instructed Noah to take seven males and seven females

of each species of the clean beasts, and two males and two females of each species of the unclean beasts into the ark. The unclean beasts were those animals that were hybrids! God considered the hybrid offspring of "finished and fixed" species to be unclean because the characteristics of those offspring did not occur as a direct part of God's creative work. Noah was also instructed to take seven males and seven females of every bird species into the ark. Not all birds were clean, but God preserved seven males and seven females of every clean and unclean bird species to help spread seeds, so as to replenish the earth with vegetation. God then predicted that in seven days hence, torrential rain would begin to fall. Noah was six hundred years old when that worldwide deluge began.

7. And Noah went in, and his sons, and his wife, and his sons' wives with him, into the ark, because of the waters of the flood.
8. Of clean beasts, and of beasts that are not that creepeth upon the earth,
9. There went in two and two unto Noah into the ark, the male and the female, as God had commanded Noah.
10. And it came to pass after seven days, that the waters of the flood were upon the earth.

That which God had predicted had come to pass. Torrential rain began to fall upon the earth. The waters gathered quickly, and the Noah family entered into the ark of safety. Inside with them was a remnant of every sort of dry land life that existed at that time, preserved for the replenishment of the earth, excepting of course, the soulless seed of mankind and their offspring that were begotten by the Sons of God. Those offspring were the "corruption" of the earth.

11. In the six hundredth year of Noah's life, in the second month, the seventeenth day of the month, the same day were all the fountains of the great deep broken up, and the windows of heaven were opened.
12. And the rain was upon the earth forty days and forty nights.
13. In the selfsame day entered Noah, and Shem, and Ham, and Japheth, the sons of Noah, and Noah's wife, and the three wives of his sons with them, into the ark;
14. They, and every beast after his kind, and all the cattle after their kind, and every creeping thing that creepeth upon the earth after his kind, and every fowl after his kind, every bird of every sort.
15. And they went in unto Noah into the ark, two and two of all flesh, wherein is the breath of life.
16. And they that went in, went in male and female of all flesh, as God had commanded him: and the Lord shut him in.
17. And the flood was forty days upon the earth; and the waters increased, and bare up the ark, and it was lift up above the earth.
18. And the waters prevailed, and were increased greatly upon the earth; and the ark went upon the face of the waters.
19. And the waters prevailed exceedingly

20. upon the earth; and all the high hills, that were under the whole heaven, were covered.
20. Fifteen cubits upward did the waters prevail; and the mountains were covered.
21. And all flesh died that moved upon the earth, both of fowl, and of cattle, and of beast, and of every creeping thing that creepeth upon the earth, and every man.
22. All in whose nostrils was the breath of life, of all that was in the dry land, died.
23. And every living substance was destroyed which was upon the face of the ground, both man, and cattle, and the creeping things, and the fowl of the heaven; and they were destroyed from the earth: and Noah only remained alive, and they that were with him in the ark.
24. And the waters prevailed upon the earth an hundred and fifty days.

Waters held in reserve in the heavens were unleashed upon the earth for forty literal days and nights. The Lord shut in eight people of the soul seed of mankind, together with a remnant of every existing kind of life that God had foreordained to replenish the earth. They were the dry land remnant from which present day life descended. Plant and sea life survived either in the floodwaters or by the remains of their seeds and eggs. Every high mountain was covered and the waters remained on the earth for one hundred and fifty literal days and nights.

The death of Cain was truly avenged "sevenfold," which means "complete." Lamech, who was a corrupt seed, killed Cain! The flood killed the tainted Sons of God, destroyed the soulless seed of mankind, but utterly destroyed the corrupt seed begotten by them. The sevenfold avengement: the corrupt seed is not only gone forever from this earth, but the nature of them has been hidden from us! The word of God mentions Cain's death being avenged sevenfold, because God realized the fallen state of man and his capabilities. Consequently, God's avengement was full and complete. There is documentation that Noah's ark still remains near the top of Mt. Ararat, Turkey, at about the "14 thousand foot level."[16]

Epilogue

Scientific facts correlate the word of God, for God is the author of truth, wherever it may be found. However, the word of God is the measuring rod of correction, having precedence over the word of science and over every other word: "... let God be true but every man a liar ..." (Romans 3:4). Therefore, facts revealed by scientific inquiries are of God, even when our interpretations of Scripture contradict those facts. That is why Scripture should be used as a guide in seeking truth because:

> All scripture is given by inspiration of God, and is profitable for doctrine, for reproof, for correction, for instruction in righteousness: That the man of God may be perfect, thoroughly furnished unto all good work (II Timothy 3:16,17).

Faith emanated from God as He created. Faith is a vital building block of accomplishment, for God said: "Without faith it is impossible to please him ..." (Hebrew 11:6). God has given to every person a measure of faith. To disregard God with that measure of faith is inexcusable:

> For the invisible things of him from the creation of the world are clearly seen, being understood by the things that are made, even his eternal power and Godhead; so that they are without excuse (Romans 1:20).

When faith becomes low-ebbed, a pressing to the word of God is needed: "So then faith cometh by hearing and hearing by the word of God" (Romans 10:17). Christians should not be piously repulsed by the fact that Adam was born through the uterus of the soulless seed of mankind, for this fact does not in any way contradict the teaching that Adam was made from dust. The laws of genetics prove that life can come only from a preceding life. God could have created Adam instantaneously from dust, but that way of creating would have been inconsistent with the way the rest of life is propagated. A mature tree created without rings would be an odd creation, as all mature trees have rings, and rings mean time. Scripture relates that all natural life was made from dust, yet we know that we were born from our parents who preceded us. To have created natural life from dust, by the creative process of evolution, was orderly and consistent. To our senses it is awesome and full of understanding!

Since the great flood in Noah's day, all people have been born through the uterus of the soul seed of mankind, including Jesus Christ. God created the natural part of man and every other kind of earthly life from earthly dust by an evolving process that took several billion years:

> And so it is written, the first man Adam was made a living soul; the last Adam was made a quickening spirit. Howbeit that was not first which is spiritual,

but that which is natural; and afterward that which is spiritual. The first man is of the earth, earthy: the second man is the Lord from heaven. As is the earthy, such are they also that are earthy: and as is the heavenly, such are they also that are heavenly. And as we have borne the image of the earthy, we shall also bear the image of the heavenly (I Corinthians 15:45-49).

God did not send Jesus to condemn the world, for it is not the fault of anyone for having inherited the fallen nature from our first parents. God's concern is to save mankind from the fall! We know now that to do evil is "exceeding sinful," for God sent the law to show us, but Adam and Eve did not have the law to show them. Before the law was given, men of the soul seed of mankind could not repent of their sins:

> ...Esau, who for one morsel of meat sold his birthright. For ye know how that afterward, when he would have inherited the blessing, he was rejected: for he found no *place of repentance,* though he sought it carefully with tears (Hebrews 12:15-17).

Death reigned from Adam, until the giving of the law to Moses, because like Esau, no one during that time could repent from the Dark Kingdom:

> Nevertheless death reigned from Adam to Moses, even over them that had not sinned after the similitude of Adam's transgression, who is the figure of him that was to come (Romans 5:14).

After Adam fell, the Dark Kingdom predominated. Only the second Adam, Jesus Christ, could cleanse sin from the soul. Before the law, anyone who came into the knowledge of good and evil moved automatically into the Dark Kingdom and could not repent. By covenants, God established two dispensations of time to provide a "place of repentance": (1) the Day of the Law which magnified sin and showed that we had need of a Savior, whom God promised to send and (2) the Day of Grace, which revealed the resurrected Savior and the Kingdom of God. Therefore, when the Dispensation of the Law (which was followed by the Dispensation of Grace) was in effect, the Dark Kingdom could no longer predominate because many souls were rescued, for the Dispensation of the Law (and of Grace) provided a "place of repentance."

The law was not given to the nation of Israel, until Moses. The souls who knew good and evil and died before the law was given, and also the Gentiles who knew good and evil and died before their coming in under the Dispensation of Grace (Acts 10), went into a spiritual Prison called "Abraham's Bosom," which is the same as Hebrew "Sheol" or Greek "Hades." So-called after Abraham, who is the "father of many nations," (for there are many nations represented in Abraham's Bosom), because: "… Abraham believed God and it was counted unto him for righteousness" (Romans 4:3). Since God did not deal with the Gentiles on a covenant basis before the Day of Grace, it is therefore unreasonable to think that He would let them go into Hell without a chance to make a choice about their hereafter life! Those Gentiles and the prelaw souls heard the gospel preached by Christ's Spirit while His natural body was in the grave. Accountable Gentiles could not enter into the Prison after Christ's resurrection! If any died between the resurrection and Acts 10, which was about seven years; they were probably lost. However, we can be assured of God's omniscience for perfect judgment. God being a loving God assures us that He preserved the lives of those Gentiles who had a conscience towards Him in that period of time.

God told Abraham that when he died, he would go and be with his fathers in peace: "And thou shalt go to thy fathers in peace; thou shalt be buried in a good old age" (Genesis 15:15). This verse reveals that Abraham's ancestors were already in the Prison place called "Abraham's Bosom," including his father Terah, who was considered unrighteous because he had been an idol worshiper from Ur of the Chaldees in Babylonia:

> And Joshua said unto all the people, Thus saith the Lord God of Israel, Your fathers dwelt on the other side of the flood in old time, even Terah, the father of Abraham, and the father of Nachor: and they served other gods (Joshua 24:2).

Therefore, the spirits of accountable prelaw people dwell in Abraham's Bosom (Sheol, Hades, or Prison), along with the spirits of accountable Gentile people who were called "strangers" or "heathens" while the Dispensation of the Law was in effect.

Scripture says that sins were not laid to the charge of those souls who died before the law was given, even though they knew and did both good and evil: "For until the law sin was in the world: but sin is not imputed when there is no law" (Romans 5:13). Before the law was given, there was no shedding of blood for the remission of sin. There were only sacrificial, burnt offerings that were not blood coverings. God cannot look upon or accept sin, so the unpurged souls could not dwell in the heavenly realm of God's light. Neither would God allow them to go into the lake of fire before the gospel was preached to them that they might be judged for either accepting or rejecting the Good News. Abraham's Bosom (Sheol or Hades) is a heavenly realm, at times called the "nether world," and a keeping place (Prison) for those departed souls, until the Day of Judgment (I Peter 4:5,6).

Jesus, the only begotten Son of the Father, was conceived in the body of a virgin, by the power of the Holy Ghost, for God developed His natural, mortal body through the only needed earthly instrument—a human female body. Even though Jesus is a God knowing good and evil (Genesis 3:22), He was never in the Dark Kingdom (that is, until He took our sins upon Himself). His Spiritual body was not inherited from Adam and Eve, but pre-existed in Heaven with His Father. Therefore, Jesus was born human and divine, because He possessed both a natural body and a pre-existing, spiritual body: "Who did no sin, neither was guile found in his mouth" (Peter 2:22).

Every soul descended from Adam has sinned, except for innocent ones who die before knowing good and evil. They too inherit the fall from Adam, but pay the penalty of the fall by the early death of their natural bodies. Their innocent spirits go straight into Heaven and dwell with God. Jesus said: "… Suffer the little children to come unto me, and forbid them not: for of such *is* the kingdom of God" (Mark 10:14). When Jesus spoke those words, the spirits of innocent children were the only spirits from this world in the Kingdom of God; for Jesus had not yet died for our sins, and no one else was clean enough to see God. Jesus then said: "Verily I say unto you, Whosoever shall not receive the kingdom of God as a little child, he shall not enter therein" (Mark 10: 15). In this verse Jesus verifies the fact that anyone who knows good and evil must become innocent and clean again, as a little child, by the rebirth of the spirit that dwells within them. Only then will they be able to receive the Kingdom of God.

After Jesus died, His natural body remained in the tomb, while His spiritual body went into three different realms: (1) He descended into Hell and unlocked the chains of death that would have held bodies to their graves forever: "… his soul was not left in hell, neither his flesh did see corruption" (Acts 2:31).

> I am he that liveth, and was dead; and, behold, I am alive for evermore, A-men; and have the keys of hell and of death (Revelation 1:18).

Translations that render Sheol or Hades as Hell do so erroneously, as "Gehenna" was used to designate the place of Hell in Old Testament times. (2) The Spirit of Jesus Christ also went into Paradise, as He promised the repentant thief who hung on a cross next to Him on that crucifixion day: "And Jesus said unto him, Verily I say unto thee, Today shalt thou be with me in paradise" (Luke 23:43). Paradise, before Christ's death, was a temporary dwelling place, but only for the spirits of those righteous Jews who had been subject to the law, under the Dispensation of the Law, and whose flesh had been purged with the blood of animal sacrifices. The thieves were crucified during a time of transition. Jesus actually died before the thieves (John 19:32,33), so the repentant thief did not need the blood of animal sacrifices to justify him for Paradise, for Christ's shed blood would also enable him to see God. Both of the thieves were Jews because only Jews could enter Paradise or Hell at that time. After Jesus died, Gentiles could neither come under grace, until the Acts 10, nor enter Abraham's Bosom. Jesus came, died, resurrected, ascended into heaven, and sent the Holy Spirit to the Jews only on Pentecost, before the first Gentile was saved. Jesus said: "… I am not sent but to the lost sheep of the house of Israel" (Matthew 15:24). Consequently, if Jesus was not sent but to the Jews, then Gentiles could not, before the conversion of Cornelius, enter into Paradise or God's Abode. The spirits of the dead, unrighteous, unpurged Jews, of course, went into the lake of fire called Hell or Gehenna. Paradise had a heavenly veil (symbolically Christ's flesh) that separated the souls that dwelled there from the face of God, until Christ purged the heavenly "things" with His shed blood:

> And almost all *things* are by the law purged with blood; and without shedding of blood is no remission. It was therefore necessary that the patterns of *things* in the heavens should be purified with these; but the heavenly *things* themselves with better sacrifices than these (Hebrews 9:22,23).
>
> For it is not possible that the blood of bulls and of goats should take away sins (Hebrews 10:4).

These evidences make it clear that the righteous Jews who waited in Paradise were not perfectly purged, until after the sacrificial shedding of the blood of Jesus Christ. They had lived under the law, but the law could not make any person fully justified in the sight of God: "Therefore by the deeds of the law there shall no flesh be justified in his sight: for by the law is the knowledge of sin" (Romans 3:20). After Christ died, both the temple and the heavenly veil were rent, for Christ purified those righteous Jews and enabled them to see God. That is why that in the Day of Grace there is no veiled division between Paradise and God's Heavenly Abode. (3) Christ's Spiritual body also went into Abraham's Bosom (Sheol or Hades) where He preached the gospel to the imprisoned souls:

> For Christ also hath once suffered for sins, the just for the unjust, that he might bring us to God, being put to death in the flesh, but quickened by the Spirit: By which also he went and preached unto the spirits in *prison* (I Peter 3:18,19).

To clarify even more, the spirits in Abraham's Bosom (Prison) had the gospel preached to them by the Spirit of Jesus Christ. He will judge them in the end as though they were living in the flesh, but who are actually living only in the spirit:

> ... For this cause was the gospel preached also to them that are dead, that they might be judged according to men in the flesh, but live according to God in the spirit (I Peter 4:6).

On the day of Pentecost, God made grace available, but only to the nation of Israel. Grace was made available to other nations of the world at a later time when God manifested it to Peter in a vision in Acts 10. Gentile nations that existed during the Dispensation of the Law were not subject to the law. Neither Israel nor the Gentile nations are meant to be under the law during the Dispensation of Grace, for God, by grace, has provided a "place of repentance" for all:

> Forasmuch then as we are the offspring of God, we ought not to think that the Godhead is like unto gold, or silver, or stone, graven by art and man's

> devise. And the times of this ignorance God winked at; but now commandeth *all* men every where to repent (Acts 17:29,30).

The two verses above reveal to us that Gentile nations worshipped idols during the Dispensation of the Law, but God winked at (excused) that ignorance, because God did not have a covenant with any Gentile people during that time. Therefore, they did not have a "place of repentance." The latter verse plainly states that God commands both Jew and Gentile people to repent under grace, which is a "place of repentance," provided for all. This last dispensation of time is the only allotted time left for repentance! Otherwise, God would not have commanded "all" to repent under this dispensation. Therefore, the verse Acts 17:30 reproves the concept of premillennialism!

The Day of Grace, which came in at the time of Christ's crucifixion did not do away with the law, only the dispensing of it, "For Christ is the end of the law for righteousness to every one that believeth" (Romans 10:4). That is why both Jews and Gentiles in the Day of Grace dwell under the law when they do not accept the sacrificial, shed blood of Jesus Christ, which results in the rebirth of the spirit: "Marvel not that I said unto thee, Ye must be born again" (John 3:7). Therefore, in the Day of Grace, both Jews and Gentiles who know good and evil, but had not been born-again, are under the law of sin and death. If they die under the law, which is now a curse, they will go into the lake of fire because grace is available to both!

The nation of Israel wandered in the wilderness for forty years, and the second generation crossed over into the Promise Land. Throughout the Old Testament, we read of Israel's history; how there were many "holy wars" fought by those chosen people of God. It was not a contradiction when God gave the commandment: "Thou shalt not kill," and then commanded them to conquer many nations, kill their inhabitants, take their possessions, and have no dealings with the Gentiles or strangers whatsoever:

> When the Lord thy God shall bring thee into the land whither thou goest to possess it, and hath cast out many nations before thee, the Hittites, and the Girgashites, and the Amorites, and the Canannites, and the Perizzites, and the Hivites, and the Jebusites, seven nations greater and mightier than thou; And when the Lord thy God shall deliver them before thee; thou shalt smite them, and utterly destroy them; thou shalt make no covenant with them, nor show mercy unto them: Neither shalt thou make marriages with them; thy daughter thou shalt not give unto his son, nor his daughter shalt thou take unto thy son. For they will turn away thy son from following me, that they may serve other gods: so will the anger of the Lord be kindled against you, and destroy thee suddenly (Deuteronomy 7:1-4).

God cannot contradict himself! It is a sin to kill humans, unless ordained of God. He loves all people equally, but deals with them in different ways at different times, not willing that any should enter into the lake of fire. In Moses' day, God chose to have a covenant relationship with the nation of Israel instead of with all nations. God's power could not have been fully seen and understood, unless a distinction had been made on a covenant basis between a people who were and a people who were not His chosen ones. God's power was shown through the Jewish people and the law, under the Dispensation of the Law, so that the entire world would understand that sin is "exceeding sinful." God's purpose would have been misunderstood if He had shown His power through all nations at the same time. We would have been "unable to see the forest for the trees" so to speak. God certainly could have chosen any nation to be His special people, but there is a reason why He chose Israel. Max R. Gaulke wrote: "God picked up and exalted Israel to a covenant status because she was the least of nations ...."[17] God chose Israel because of its size. It was the smallest nation at that time. Other nations that were larger were meant to be impressed by Israel's power, believing that Israel's God must be the true God. He did not choose her because He loved her above other nations. He loved her because He chose her! Similarly, God did not really "hate" Esau. "As it is written, Jacob have I loved, but Esau have I hated" (Romans 9:13). That expression was used to show His acceptance of Jacob over Esau, according to the promise, because Esau represented the state of not being under the promise. Since we are no longer living under the Dispensation of the Law, then God's power and might is no longer shown through the literal nation of Israel. God's power and might, since Pentecost, is shown through the redeemed— a spiritual, holy nation, called "Israel"— not the literal nation that uses the same name.

The Israelites were expected to walk in faith and obedience before God. Under the Mosaic covenant, God forbade them to mix with or marry Gentiles. He did not want His chosen ones to be exposed to the worship of false gods. However, at times, some disobeyed. King David's great-grandmother, Ruth, of the Bible book of Ruth, was a Gentile woman. His great-grandfather Boaz, a Jew, married her. God's blessings could reach Ruth because she had a conscience towards the true God, whereby she would not influence others to idol worship. However, Ruth could not be judged by the Law of Moses when she died because she did not have any Jewish blood in her. God's covenant was with the Israelites, and she was a Moabite. Ruth's spirit dwells in Abraham's Bosom (Shoel, Hades, or Prison) where she heard the gospel preached by the Spirit of Christ. She will be judged by that gospel on the Day of Judgment.

There are two reasons why "death reigned from Adam to Moses" (Romans 5:14): (1) Before Moses, no one knew that sin was "exceeding sinful," for the law had not yet been given. Everyone knew during that time that their natural bodies would die sooner or later, whether or not they committed a sin, but they were ignorant of the whole truth: "... Nay, I had not known sin, but by the law ... For without the law sin was dead" (Romans 7:7,8). Sin was dead, but God could not allow spiritual bodies of accountable people at that time to dwell in the same Heavenly Abode where He and the spirits of innocent ones dwelled, so they dwelled in Abraham's Bosom. (2) Before Christ died and resurrected, Satan had the "creature" (natural body) bound to the grave: "For the earnest expectation of the creature waiteth for the manifestation of the Sons of God" (Romans 8:19). This reveals that the dead would not be able to resurrect until men could once again be called "Sons of God." Before the law, men were called "Sons of God" because their sins were not imputed to them. After the law was given, men lost the right to that name, because the law condemned them. Even the bloody sacrifices of bulls and goats could not cleanse them enough for that privileged name, so God called His chosen ones "Servants of God." Under the Dispensation of the Law, righteous Jews, as servants, sought salvation by faith and works. Under the Dispensation of Grace, Christ gave believing Jews and Gentiles the power to become "Sons of God," when they are perfectly purged by the blood of Christ:

> But when the fulness of the time was come, God sent forth his Son, made of a woman, made under the law, To redeem them that were under the law, that we might receive the adoption of sons. And because ye are sons, God hath sent forth the Spirit of his Son into your hearts, crying, Abba, Father. Wherefore thou art no more a servant, but a son; and if a son, then an heir of God through Christ. (Galatians 4:4-7).

About 14 hundred years were spent under the Dispensation of the Law, which convinced the world that a Savior was needed. After Jesus came and filled that need, there was no longer a covenant for just one literal nation. Instead, there was a new covenant for a spiritual nation of people, out of every nation. In fact, the spiritual Jewish nation had the new covenant for about seven years before the Gentiles entered into it. God brought in people from Gentile nations to make the Jews jealous, that He might save some of them. (Romans 11:11). Therefore, some of them have been saved: "… there is a remnant according to the election of grace" (Romans 11:5). On the Day of Pentecost, which was ten days after Jesus ascended into heaven, three thousand Jews were saved. It was the birth of the true church—The Bride of Christ! They were grafted back in, individually, for they were the natural branches and the first to be saved under the Dispensation of Grace. Jesus said: "I am not sent but unto the lost sheep of the house of Israel" (Matthew 15:24). Jesus was born, crucified, died, resurrected, and ascended into heaven before God brought the Gentiles under the new covenant. To believe salvation for the literal Jewish nation in a future, millennial age is contrary to the new covenant that is by grace. A premillennial belief is professing that what Jesus did for the unsaved nation of Jews at His crucifixion and at Pentecost was not enough. As Jesus was expiring, He said: "it is finished" (John 19:30). What Jesus did to save mankind is not only over, it is sufficient!

The wild branches of the Gentile nations were grafted in under the new covenant, beginning with Cornelius, an Italian, about seven years after Pentecost (Acts 10). It is false teaching to say redeemed Jews are not part of the church! Gentiles are joined into redemption with them, into their true church, and into their new holy nation that was ordained by God through Jesus on Pentecost. We are not only "Sons of God," but also "kings, priests, and spiritual Jews" (Revelation 5:10):

> For he is not a Jew, which is one outwardly; neither is that circumcision, which is outward in the flesh: But he is a Jew, which is one inwardly; and circumcision is that of the heart, in the spirit, and not in the letter; whose praise is not of men, but of God (Romans 2:28,29).

Therefore, we can see why God had to reveal the law before He could reveal Christ. We cannot realize our need of a Savior until the law reveals to us our sins! Before Moses, neither the law nor grace had been revealed, so "death reigned" because there was no "place of repentance."

During the Dispensation of the Law, some Gentile people realized that Israel's God was the true God. Therefore, by faith they kept the law by nature (conscience, or the law written in their hearts), but not by letter. To keep the law by nature was accounted to them for some justification in the natural realm, even though they were not of the circumcision, for God had no covenant with them. There were sins on their souls from the old sin nature, which had not been purged with blood. After they died, their spiritual bodies went into Abraham's Bosom (Sheol, Hades, or Prison)(Romans 2:11-16).

Gentiles were not required by God to keep the letter of the law. Offerings of blood sacrifices and circumcision of the flesh had no effect on them spiritually, for the law was not given to them. God does not recognize such rituals, even by the Jews, in the Day of Grace.

Both conscientious and unconscientious Gentiles who died during the Dispensation of the Law, rest in the Bosom of Abraham (Sheol, Hades, or Prison) until the Day of Judgment. Abraham's Bosom is not in the higher realm of Heaven where God dwells, but it is in Heaven. In the word of God we read of a poor beggar named Lazarus who begged for the crumbs that fell from a rich man's table. The beggar died and was carried by angels into Abraham's Bosom, and the unrighteous rich man died and went into the lake of fire called "Gehenna" or "Hell" (Luke 16:19-26). Lazarus was a Gentile and the rich man was a Jew. During the Dispensation of the Law, only righteous Jews went into Paradise. In this Biblical account, the rich man addressed Abraham as Father. God would not allow Abraham or anyone else to usurp His authority or the name "Father" if they were dwelling in His Abode. Jews were not able to call God "Father" under the Dispensation of the Law, for they were not Sons, but servants; so too, Abraham's Bosom cannot be in the same realm where God dwells.

In the word of God we also read about a harlot name Rahab. Her natural life and the natural lives of her relatives were spared from the Jews who captured their city. Rahab was a Gentile who had a conscience towards God, thereby making her justified to receive God's blessings in this natural realm (Joshua 2:1-24 and Joshua 6:16-25). Her spirit and the spirits of her relatives are also in Abraham's Bosom (Shoel, Hades, or Prison) awaiting the judgment.

"It is a fearful thing to fall into the hands of the living God" (Hebrews 10:31). The Jews as a nation were God's chosen people under the Dispensation of the Law, but the Jews were also the first to enter into the lake of fire known as Gehenna or Hell: "... For unto whomsoever much is given, of him shall be much required ..." (Luke 12:48); for they were the only ones to whom God committed His oracles. However, as a nation they did not keep His oracles, but disobeyed; so the literal Jewish nation was "broken off" as being a holy, chosen nation of God. That is known as "The Time of Jacob's Trouble" because it was the time that the literal tribes of Israel (descended from Jacob) were dissipated forever (Jeremiah 30:7). God had a plan that could save them, in a spiritual way, in spite of the "trouble," as Jeremiah prophetically declared: "... in the latter days ye shall consider it" (Jeremiah 30:24). The Dispensation of Grace, which is the latter day, and the "grafting in" of the Gentile peoples, brought about the end of the Dispensation of the Law and Israel's dominion over other nations in the eyes of God. In the Day of Grace, God does not have a literal chosen nation, but a spiritual chosen nation—the redeemed of God (I Peter 2:9)! The Jews being a natural branch may be grafted back in (as a redeemed people; not as a literal holy nation unto itself), but like Gentile peoples, only by grace, which is by accepting Jesus Christ as Savior. Therefore, all the redeemed of God are spiritual Jews:

> Thou wilt say then, The branches were broken off, that I might be grafted in. Well; because of unbelief they were *broken off*, and thou standest by faith …. And they also, if they abide not still in unbelief, shall be *grafted in*: for God is able to graft them in again (Romans 11:19-23).

In the day of grace all the redeemed are Israel! The unredeemed of both Jews and Gentiles can still be grafted in, by turning to Jesus as Savior, and become true Jews in the "Kingdom of God." In the Day of Grace, the children of the promise are by faith in Christ (in which good works follow), but not by works of the law. It is therefore no longer a literal issue on how Jews live under the law, but a spiritual issue on how redeemed Jews and Gentiles live under grace:

> There is neither Jew nor Greek, there is neither bond nor free, there is neither male nor female: for ye are all one in Christ Jesus. And if ye be Christ's, then are ye Abraham's seed, and heirs according to the promise (Galations 3:28,29).

When Christ Jesus said: "… I am not sent but unto the lost sheep of the house of Israel" (Matthew 15:24), He was revealing that the new covenant was primarily meant for literal Israel. The following verses of Scripture was directed to the nation Israel, as God was establishing the "new covenant" with them:

> For if that first covenant had been faultless, then should no place have been sought for the second. For finding fault with them, he saith, Behold, the days come, saith the Lord, when I will make a *new*

> *covenant* with the house of Israel and with the house of Judah: Not according to the covenant that I made with their fathers in the day when I took them by the hand to lead them out of the land of Egypt; because they continued not in my covenant, and I regarded them not, saith the Lord. For this is the covenant I will make with the house of Israel after those days, saith the Lord; I will put my laws into their mind, and write them in their hearts: and I will be to them a God, and they shall be to me a people: And they shall not teach every man his neighbour, and every man his brother, saying, Know the Lord: for all shall know me, from the least to the greatest. For I will be merciful to their unrighteousness, and their sins and their iniquities will I remember no more. In that he saith, A *new covenant*, he hath made the first old. Now that which decayeth and waxeth old is ready to vanish away (Hebrews 8:7-13).

Literal Israel, as a nation, failed to attain to the law of righteousness because the law of righteousness includes Jesus Christ as Messiah or Savior. God cast away the entire Jewish nation as being a holy nation. The old covenant is no longer binding to the Jews, for it will never again be restored. Since the Acts 10, the new covenant is for all people if they will accept it. That is why the law condemns when grace is rejected.

Before Jesus ascended up to Heaven, He assembled together with His believers and spoke about the Kingdom of God. With hindsight, we now know that Pentecost was a time of "restoration," but not for the literal nation of Israel, unless *all* would have repented.

> When they therefore were come together, they asked of him, Saying, Lord, wilt thou at this time restore again the kingdom to Israel (Acts 1:6)?

This Bible verse is often used to give credence to the false teaching of premillennialism, because many believed then and still do now that God would restore the kingdom to literal Israel. He would not have cut them off if that was His intention. He gave them many opportunities and times to repent. The spiritual kingdom of Israel was "restored" at Pentecost, which happened before the literal destruction of the people, temple, and land by Titus.

Adam Clarke wrote that the Greek verb from which the word "restore" was translated, has also another meaning in the Greek language, such as "ending," "abolishing," "blotting out," or "making an end" of a thing. He goes on to say:

> On this interpretation the disciples may be supposed to ask, having recollected our Lord's prediction of the destruction of Jerusalem, and the whole Jewish commonwealth, Lord Wilt thou at this time destroy the Jewish commonwealth, which opposes thy truth, that thy kingdom may be set up over all the land? This interpretation agrees well with all the parts of our Lord's answer, and with all circumstances of the disciples, of time, and of place.[18]

Since the Greek word for "restore" can also mean "abolish," it seems that it would not matter which way the disciples were asking because both events came to pass. Pentecost was a time of restoration, and Titus was a time of abolishment!

It is inconceivable to think that premillennial believers overlook a part of history, when making interpretations of end time prophecies. The "millennium" is symbolic of "one full day of grace." Nebachadnezzar's dream man is symbolic of kingdoms that have transpired in times past (Daniel 2). His ten toes are not a symbol of a future ten-nation confederation. They represent the German tribes that invaded the Roman Empire, which helped cause it to fall. Their symbolic short reign is prophesied in the following verse:

> And the ten horns which thou sawest are ten kings, which have received no kingdom as yet; but receive power as kings one hour with the beast (Revelation 17:12).

Historic chronology should never be overlooked! It must be emphasized that it is inexcusable to use another event to replace a part of history, and project it into the future. In this account, replacing a part of history with a future non-Biblical idea distorts the dream man.

In Acts 1:6, the disciples asked Jesus: "… Lord, wilt thou at this time restore again the kingdom to Israel?" This question verifies the fact that the nation of Israel had been "broken off." God had sent many prophets to warn them, and He had given them much time to repent, but they would not do so. There were some Jews in that nation who were righteous, and they, too, were broken off with the rest before Pentecost because they were part of that nation. Many disciples then and now believe God will restore the literal kingdom of Israel in the future, after the rapture of the saints. No doubt many misinterpretations occur at this conjecture! In Acts 1:7,8, Jesus tells them that only God knew when the restoring would occur, but Jesus knew how and why it would happen, and the restoring did occur on the Day of Pentecost. In the Day of Grace, which is called "The Last Days," everyone can have their own Pentecost. "And it shall come to pass in the last days, saith God, I will pour out of my Spirit upon all flesh ..." (Acts 2:17). This is God's endeavor in seeking to save accountable people, which started on Pentecost, and will continue until the end of the Day of Grace.

Preaching and teaching premillennialism does a grievous injustice to those Jews that are not born-again. It gives them a false hope that God will not judge them, because they teach them that they are still a holy, chosen nation.

God broke off the literal, nation of Israel. He replaced it with a spiritual, holy nation of redeemed people, from every literal nation. Therefore, it is illogical to have two holy nations at the same time. There are only two dispensations of time when God showed redeeming power for saving mankind, and He did that with covenants: (1) The Covenant of the Law and (2) The Covenant of Grace. The Bible focuses on these two dispensations of time as the only time anyone can be saved. The righteous Jews, who died under the Dispensation of the Law and were already in Paradise, were perfectly cleansed when the Spirit of Christ went there while His physical body was in the grave.

When Jesus was teaching His disciples to pray, He taught them "The Lord's Prayer" (Luke 11:1). "... Thy Kingdom come, thy will be done ..." is part of that prayer and was prayed before His crucifixion and ascension. The prayer was answered ten days after the ascension, on the Day of Pentecost. Since God's will has been "done on earth, as in heaven," then that part of the prayer said in our time, after Pentecost, is prayed in vain because the Kingdom of God has been on earth since Pentecost.

Jesus said: "... behold the kingdom of God is within you" (Luke 17:21).

> And he said unto them, Verily I say unto you, That there be some of them that stand here, which will not taste of death, till they have seen the kingdom of God come with power (Mark 9:1).

Since Pentecost occurred only ten days after the ascension, then Jesus knew that some of them who were standing there would still be living after such a short time. Those Jews saved on Pentecost and since then are Jewish Christians. The Gentiles saved at the time of Acts 10 and since then are Christian Jews. All the saved are called "Israel." Literal unsaved Jews are not to refer to themselves as "Jews" or "Israel." Jesus spoke to the church (redeemed) in Smyrna: "... I know the blasphemy of them which say they are Jews, and are not, but are the synagogue of Satan" (Revelation 2:9).

Many students, teachers, and others in our society want outward prayers said in our schools. Outward prayer is sometime needed, but silent prayers can be very effective. We already know that diversity among cultures and religious beliefs can cause strife of all kinds. Kindness with love towards one another would be a better way to draw people to the true God. The sad thing is that not everyone will try to find the true way. If you are not born-again by accepting Christ as Savior, then how can your prayers reach God? There are, no doubt, many prayers routinely said that never reach God. Jesus said: "I am the way, the truth, and the life: no man cometh unto the Father, but by me" (John 14:6).

There is much controversy over The Ten Commandments being placed in public places. It is a wonderful guide for knowing what is unrighteous, if not kept. However, The Great Commandment should be placed with the Ten Commandments wherever they are displayed. The Ten Commandments cannot be kept without first keeping The Great Commandment; it shows all what is righteous if kept:

> Thou shalt love the Lord thy God with all thy heart, soul, and mind. Thou shalt love thy neighbor as thyself (Matthew 22:37-39).
>
> But I say unto you which hear, Love your enemies, do good to them that hate you (Luke 6:27).

Under the Dispensation of Grace, both unredeemed Jews and Gentiles can enter into the lake of fire when they die. Therefore, to commit an unloving act towards anyone in the Day of Grace is to break the Commandment of Love that Jesus gave to the world. It is only reasonable that Christ would give us the Commandment of Love, the *last* commandment: "For Christ is the *end* of the law for righteousness to every one that believeth" (Romans 10:4). Jesus fulfilled the law for Jesus is love. The Ten Commandments cannot be broken if the Commandment of *Love* is kept:

1. If you love God, you will not have other gods.
2. If you love God, you will not make a graven image to worship.
3. If you love God, you will not profane His name.
4. If you love God and neighbor, you will keep the Sabbath Day holy, by keeping

all days holy with love.
5. If you love your parents, you will honor them.
6. If you love all, you will not murder anyone.
7. If you love all, you will not commit adultery with anyone.
8. If you love all, you will not steal from anyone.
9. If you love all, you will not lie to anyone.
10. If you love all, you will not covet the mate or possessions of anyone.

God does not ordain "holy wars" in the Day of Grace! How could he? In the Day of Grace, God's redeemed come from every nation, race, and tongue, and God would not ordain the Sons of God to fight one another. God allowed Israel's dominion over other nations under the Dispensation of the Law, especially when they were obedient. Under the Dispensation of Grace there is no literal nation that has dominion over any other nation in the eyes of God. Under the Dispensation of the Law, Gentiles who knew good and evil went into Abraham's Bosom (Shoel, Hades, or Prison) when they died. They wait there peaceably for the judgment. God allowed the death of their natural bodies to be taken for Israel's sake, for the spirits of Gentiles could not be judged at that time.

The Commandment of Love could not be kept if God ordained "holy wars" in the Day of Grace. Under the Dispensation of Grace, all nations have the God-given right, as higher powers, to enforce "godly" laws (Romans 13). Therefore, peaceful nations may have to war against other nations that do evil deeds, but evil deeds that cause wars in the Day of Grace are ordained of Satan, not of God.

There are two kingdoms in the earth: (1) The Kingdom of Light and (2) The Kingdom of Darkness. Citizens of each kingdom may sometime feel scorn, when being imposed upon by the other. God left the law in the earth for the lawbreaker. Therefore, everyone must judge himself as to which kingdom he resides, but God is the ultimate judge:

> ... The law is good, if a man use it lawfully; Knowing this, that the law is not made for a righteous man, but for the lawless and disobedient, for the ungodly and for sinners, for unholy and profane, for murderers of fathers and murderers of mothers, for manslayers, For whoremongers, for them that defile themselves with mankind, for menstealers, for liars, for perjured persons, and if there be any other thing that is contrary to sound doctrine ... (I Timothy 1:8,9).

God did not do away with the law, but fulfilled it with grace. Fallen mankind remains under the law until being born-again. Anyone living under the law in the Day of Grace is living under a curse, because the law cannot save you. It is the power of the law that condemns one to hell. An unsaved couple getting married is recognized by God as being "joined together" by the law. An unsaved couple getting divorced is recognized by God as being "divorced" by the law. A born-again couple (male and female) getting married by the law is also "joined together" by the law and God, but should not be "put asunder" by the law. If you become born-again after a lawful marriage, then your marriage should, by consent, move under grace. Only God knows how many marriages have been "joined together" by His law and His grace:

> ... What therefore God hath joined together, let not man put asunder (Matthew 19:6).

> ... If righteousness come by the law, then Christ is dead in vain (Galatians 2:21).

In the book of Ezekiel, God made a Covenant of Peace so that there would be a restoration of Israel and Judah. He told Ezekiel to write their names, "Beauty and Bands," upon two sticks that He would "join together" and become "one in His hand" (Ezekiel 37:19). God desired the reunion of those two kingdoms, when the literal, Jewish nation would accept Jesus on Pentecost as Messiah, the Son of God. However, God broke that covenant before the birth of Jesus:

> ... I took my staff, even Beauty, and cut it asunder, that I might break my covenant which I made with the people (Zechariah 11:10).

> Then I cut asunder mine other staff, even Bands, that I might break the brotherhood between Judah and Israel (Zechariah 11:14).

The making of the Covenant of Peace and the prophesy of the breaking of that covenant was about 70 years apart. The prophets were sent to Israel, and they told Israel the blessings that would occur, if the nation would repent. When Christ brought in the New Covenant on Pentecost, only three thousand were "grafted in" and became "The Spiritual Holy Nation of Israel—the Bride of Christ—the True Church." Literal Israel is no longer a holy nation because the entire nation did not repent. If they had, the Gentile nations would have gone up to the literal Holy Nation of Israel to be enlightened and be saved which was Plan-A of God. However, Plan-B of God took place when the first Gentile person, named Cornelius, became saved and "grafted in" under the New Covenant. God did this to make the Jews jealous, that He "... might save some of them" (Romans 11:14). Nonetheless, God will graft each one into the Spiritual Holy Nation if they will repent and accept Jesus the Savior before His Second Advent.

During the Dispensation of the Law, Gentiles were not under any covenant. The Bible says: "And the times of this ignorance God winked at; but now commandeth *all* men every where to repent" (Acts 17:30). Therefore, it is unnecessary to relate to a futuristic reign of Christ on earth, since the two dispensations were given for "all" men to repent. This is symbolized by the vision of Ezekiel, Chapter 1: The faces of the angelic beings represent the twelve tribes of Israel that were found righteous in the Day of the Law. The wheels, and the eyes in the wheels, represent the saved of the Day of Grace, both Jews (small wheel, representing some), and Gentiles (large wheel, representing many). They follow the Holy Spirit where He leads! During the Dispensation of Grace, the redeemed of both Jew and Gentile nations are called "Israel." Therefore, all the redeemed are numbered with the symbolic one hundred forty-four thousand of the tribes of Israel (Acts 17:30). The twenty-four elders around the throne of God represent the saved of the earth: the twelve tribes of Israel under the Dispensation of the Law, and the twelve apostles, under the Dispensation of Grace.

Ezekiel was called upon to prophesy to the valley of dry bones: "... Prophesy upon these bones, and say unto them, O ye dry bones, hear the word of the Lord" (Ezekiel 37:4). In this chapter of the Bible, God is revealing to us that the valley of dry bones are the bones of those souls who accepted the gospel when Christ's Spirit preached to them in Abraham's Bosom (Sheol, Hades, or Prison). There are many nations represented in that body of souls. Those saved prisoners are also of the "Spiritual Holy Nation of Israel," for they are counted among the redeemed. They had died and heard the gospel preached by Christ before the Kingdom of God came to earth on Pentecost, therefore, the Holy Spirit had not baptized them! It must be recognized that Jesus said: "… It is expedient for you that I go away: for if I go not away, the Comforter will not come unto you …" (John 16:7). Those prisoners will remain in Abraham's Bosom until the resurrection! On that day, they will be raised from their grave, and then God will baptize them with the Holy Spirit so they can ascend with the rest of the redeemed. The resurrected—lost ones—cannot ascend because they have not the Spirit of God.

At a later time, Zechariah prophesied that Judas Iscariot, one of the apostles, would betray Jesus: "… So they weighed for my price thirty pieces of silver" (Zechariah 11:12). Jesus quoted the Old Testament as He went about teaching the Jews. He was aware of the prophecies that had been written! He knew the literal nation had been "broken off," and He called them "the lost sheep of the house of Israel." At Pentecost, the Jews were first then—not first now. At Acts 10, the Gentiles were second then—not second now. In the Day of Grace after Pentecost, any Jew or Gentile can be saved. This special dispensation gives sufficient time for every soul to repent and accept Jesus Christ as Savior. In the Day of Grace, God "… pours out His Spirit upon all flesh …" (Joel 2:28), to woo all into the Kingdom of God. Christ will not return to the dusty, bloody streets of literal Israel, because there is no need to do so. As one can well see, through present and past conditions, there is not anything holy about trying to live under the Law in the Day of Grace. The wrath of men has truly fallen upon the unsaved of literal Israel. Jesus stated: "Behold, your house is left unto you desolate" (Matthew 23:38). Jesus spoke these words before His death on the cross. He knew Plan-B would be activated, as He was aware of the words of the prophets in Jeremiah 22:5 and Zechariah 11. He was also aware that a spiritual nation of Israel would be born on Pentecost.

The numbers 666 in the Bible are a symbol of organized religions, which are religions of men. Since the redeemed of the earth are the true church, then any other literal, so-called churches have to be counterfeit and full of bondage. When discerning the true church, the expounding of Scripture by St. Paul in 52 AD enlightens us as to when and how organized religions began. "… The mystery of iniquity doth already work …" (II Thessalonians 2:7). Paul was able to see the subtle beginnings of the rise of a religious hierarchy that led to the beginning of organized religions.

In 70 AD a Roman military man named Titus, who was the son of emperor Vespasian, captured Jerusalem and destroyed the temple, thereby causing the Jewish sacrifices to end. About 312 AD Emperor Constantine made Christianity a state religion. Both historic events were caused by the power and might of the so-called Holy Roman Empire! The Bible states that the same power that took away the daily sacrifice is also the same power that set up the "abomination that maketh desolate":

> And arms shall stand on his part, and they shall pollute the sanctuary of strength, and shall take away the daily sacrifice, and they shall place the abomination that maketh desolate (Daniel 11:31).

Making Christianity a state religion was not a holy undertaking after all. Constantine set himself above God in his endeavor to make Christianity a legitimate religion, of the land, which can only be done by God. The true church is not a building, denomination, ministry, government, or any other organized, worldly group. The true church was founded by God on Pentecost and was meant for every person on earth. The true church is a people who know they are born-again by the "Spirit and Word"—the two witnesses.

In 1948, literal Israel became a state, but only by the sheer will of men. It was maneuvered by political forces, and not ordained by God to be restored to its former glory as a holy nation. In this Day of Grace, there is only one holy nation and church—the redeemed of God! Literal nations cannot become holy in the Day of Grace, because not everyone in a nation will be born-again. However, God's covenant is there for everyone, but they must accept it.

About 342 years separated the two events and persons (Titus and Constantine) who exposed "the abomination that maketh desolate," better known as "organized religions." Characteristics that helped expose the abomination of desolation:

> Neither shall he regard the God of his fathers, nor the desire of women, nor regard any god: for he shall magnify himself above all (Daniel 11:37).

During the time of the Crusades, many saints were slain because they would not embrace an organized religion posed as being Christian, but better known in Scripture as: "… the great whore that sitteth upon many waters" (Revelation 17:1), "… which are peoples, and multitudes, and nations, and tongues" (Revelation 17:15). The mother harlot represents the first counterfeit organized religion, posed as being Christian. She and many daughters born out of her have thrived and maintained constancy down through the ages by holding many souls captive! Some organized religions and cultures put women in subjugating positions. Therefore, in many places they are considered unequal and suffer abuse. Geographically speaking, the mother harlot sits on seven mountains, which is the heart of Vatican State: "... the seven heads are seven mountains, on which the woman sitteth" (Revelation 17:9). She is a church/state—the bare remains of the Unholy Roman Empire! She is "Mystery Babylon," the place where many of God's people, in the Day of Grace, are in bondage. The word "Babylon" is the name of the place where literal Israel went into bondage, when the Dispensation of the Law was in effect, but that was not a mystery. The word "mystery" here denotes spiritual bondage— not literal bondage.

Another event called the "Reformation Movement" of the 16th century was another deception of Satan that caused many to see it as a "reformation glory." The organized religions that erupted from that event may have expounded more pure doctrine, but organization is one of the places where the line is drawn when discerning the status of a religious body. Those religions actually separate the redeemed of God, which is very unholy because God ordained the redeemed to assemble together. Suffering bondage to assemble is sometime tolerated, but if God considers it a hated thing, then we should not lend credence to it. God sent the Holy Spirit to lead, guide, and comfort. Organized religions usurp Holy Spirit authority and leadership! God said: "... Come out of her, my people, that ye be not partakers of her sins, and that ye receive not of her plagues" (Revelation 18:4).

In reality, the true church and every state are already

separated! That is because the true church is the redeemed from every nation. Nations do not always have the same kind of government, and in this Day of Grace there is no literal nation that has a covenant with God. Therefore God loves all nations equally! Any literal nation can be built on Christian principles and should be, but that does not make any of them a holy nation! The redeemed of God have become a spiritual, holy nation. Therefore, organized religions are unnecessary. God does not want His redeemed to be in bondage. Jewish Christians and Christian Jews do not need to keep rituals and traditions of organized religions: "If the Son therefore shall make you free, ye shall be free indeed" (John 8:36).

Now it is profoundly stated with Biblical proof that the symbolic, one thousand-year reign of Christ on earth is nearing its end—not its beginning! The Kingdom of God (the saved people of the Day of Grace), has bound Satan during these one thousand symbolic years, by the word of God. He will be loosed for a short time, after the rapture (ascension of the redeemed), because the saints will be gone to be with Jesus (Revelation 20:7).

Since the seventh day of rest is nearing its end, as well as the one thousand symbolic years that the Kingdom of God has been reigning on earth in the bodies of the redeemed (I Corinthians 6:19), then there cannot be a literal person known as "Antichrist." There is a spirit known as such, and he is in the hearts of those who do not believe in Christ as Savior of mankind. Many souls on earth are filled with the spirit of Antichrist!

The two witnesses in Revelation are not literal men who will be raised from the dead. The witnesses are symbolic of the "Spirit and the Word." When a person becomes born-again, there is realization, by the agreement of the two witnesses, when something is true and that you are a *Son of God*. People who use the Judeo-Christian word to describe God's people should use it discriminately. It should not be used to include any literal nation; not even the literal nation of Israel!

Premillennialists who preach and teach a one thousand-year reign of Christ on earth, in the Middle East, also spread another false teaching about the "Battle of Armageddon." It is a true battle of the spirit world, and we all fight in this battle between good and evil if we live to the age of accountability. When one is tempted to sin, he is fighting in that spiritual battle—if he chooses to overcome evil with good. The Battle of Armageddon can be fought only in this Dispensation of Grace, because the Spirit of God would not enter into the souls of men until Pentecost. The Holy Spirit would fall upon men of old times, but not into them because Jesus had not died to wash away their sins. The redeemed win in this battle, called "The Battle of Armageddon," because only the redeemed have the power of God in them to help them win. This battle will be fought until the end of time.

It seems contradictory to say that we should not kill, and then God allows the higher powers that be to execute wrath upon them who do evil deeds: It is a sin to kill human beings, unless ordained of God:

> For rulers are not a terror to good works, but to the evil. Wilt thou then not be afraid of the power? do that which is good, and thou shalt have praise of the same: For he is the minister of God to thee for good. But if thou do that which is evil, be afraid; for he beareth not the sword in vain: for he is the minister of God, a revenger to execute wrath upon him that doeth evil (Romans 13:3,4).

God left the law in the earth for evildoers, because God realized that people in the Dark Kingdom would do evil here on earth, until the end time. God gives all men a chance to be born-again, but many will not turn to Him. The law is not perfect, but it helps keep chaos down for the sake of the innocent. If the ministers of the law make godly laws, then it makes for more peace in the earth. If the ministers make ungodly laws, then the innocent ones also suffer. God does not force anyone into the Kingdom of Light, but covets the salvation of everyone, including ministers of the law. He also expects them to rule with love and compassion, even when they administer punishment. Wars, death penalties, and other avengements should be kept to a low, justly minimum. Rehabilitation may help save a lost soul by giving him more time to repent. Many ungodly laws have been made down through the ages and have caused much death, grief, and despair. God's revenge will certainly fall upon those ministers who execute unjust laws and render unnecessary punishment. In the Day of Grace, one does not take an "eye for an eye, or a tooth for a tooth." Scripture exhorts:

> See that none render evil for evil unto any man; but ever follow that which is good, both among yourselves, and to all men (I Thessalonians 5:15).

Another type of bondage that organized religions put upon mankind is the teaching of "tithing." God ordained the nation of Israel to give a tenth of their first fruits to the tribes of Levi and Benjamin, since they did not receive an inheritance. Their duties were to care for the temple and government. The other tribes were required to give to the poor. There is no place under the New Covenant or in the New Testament, in this Day of

Grace, that commands tithing. In fact, there is Scripture that shows us by Jesus' teachings during the transition between dispensations that everything we possess belongs to God. The rich, young ruler who kept the Commandments and undoubtedly kept the tithe was prompted by Jesus to give 100% to the poor (Luke 18:18-22). In the Day of Grace, the Holy Spirit can help us to know where, when, and how much to give. Whether or not you give 100%, 10%, or any amount, it should be called an "offering" or "alms." In this Day of Grace, many in organized religions are made to feel guilty if they don't tithe. Some give away their sustenance to others and cause members of their own family to suffer needs. The word of God states:

> But if any provide not for his own, and specially for those of his own house, he hath denied the faith, and is worse than an infidel (I Timothy 5:8).

God works through mankind for His own purpose and glory, and for the good of mankind. We know He hardened Pharaoh's heart, but that does not reveal how Pharaoh will be judged. In Pharaoh's day the law had not yet been given:

> For the scripture saith unto Pharaoh, Even for this same purpose have I raised thee up, that I might shew my power in thee, and that my name might be declared throughout all the earth (Romans 9:17).

God used Pharaoh and others also for His own purpose and glory, but that does not mean that God covets the death of sinful people:

> ... As I live, saith the Lord God, I have no pleasure in the death of the wicked; but that the wicked turn from his way and live ...(Ezekiel 33:11).

Many times God cannot bless, for evil will cause anguish and tribulation in the earth: "Tribulation and anguish, upon every soul of man that doeth evil, of the Jew first, and also of the Gentile" (Romans 2:9). Pharaoh suffered some anguish and tribulation in the flesh, but that does not mean that his spirit went into the lake of fire when he died. Pharaoh's spirit went into a prison called Abraham's Bosom (Sheol or Hades) where he heard Christ's Spirit preach the gospel. On the Day of Judgment Pharaoh will receive eternal bliss if he received the gospel with a glad heart.

When God destroyed the cities and plains of Sodom and Gomorrha, He was not trying to wipe out sin or destroy the people because of their sins per se, for He would have failed if that was His purpose. God knows that all men (since Adam and Eve) are innocently born with sin. God sent Jesus to rescue man, not to condemn him, for only the law condemns man of sin: "For God sent not his Son into the world to condemn the world; but that the world through him might be saved" (John 3:17). Fire and brimstone fell on Sodom and Gomorrha before the law was given, so only the natural bodies of the people were destroyed. The spirits of them all went into Abraham's Bosom, unless there were innocent children present. If there were, then their spirits went into God's Heavenly Abode. The fire and brimstone were sent by God, but only as an "example" of an everlasting fire:

> Even as Sodom and Gommorha, and the cities about them in like manner, giving themselves over

to fornication, and going after strange flesh, are set forth an *example*, suffering the vengeance of eternal fire (Jude 1:7).

God, by His wrath, had to impress upon mankind the seriousness of sin and its consequences! The "example" given, of the burning of the cities and plains was to show all who lived under the Dispensation of the Law and thereafter that Hell is a place to avoid, for it burns with literal fire and brimstone:

And turning the cities of Sodom and Gomorrha into ashes condemned them with an overthrow, making them an *ensample* unto those that after should live ungodly (II Peter 2:6).

God destroyed Sodom and Gommorha by fire as an "example" of an everlasting fire, but He did not destroy the earth by water as an "example" of an everlasting flood! The Sons of God in Noah's day had sinned just as all men have sinned, but God did not destroy their natural bodies because of sin per se; for God knew sin would remain on earth until the end of time! Therefore, there had to be another purpose for the flood, and that purpose was to destroy a "corrupt" life that God had not willed to exist and the possibility of it ever existing again. The prevalence of the flood was great and long-lasting enough to erase any doubts from our minds and hearts, that what God intended to destroy, He succeeded in doing so. Therefore, man will never be able to reproduce by crossing man with other species, as to do so would cause "corruption" to occur in the earth again.

God did not leave man in a hopeless condition because Adam fell from innocence into the Dark Kingdom and caused the soul seed of mankind to inherit the fallen nature. God foreordained a way for mankind to be saved:

> For God so loved the world, that he gave his only begotten Son, that whosoever believeth in him should not perish, but have everlasting life (John 3:16).

Less than forty hours after Jesus died, His spiritual body was back with His natural body in the tomb where His glorious resurrection occurred. He was the first to experience such miraculous power, making it possible for the soul seed of mankind to do likewise someday:

> And from Jesus Christ, who is the faithful witness and the first begotten of the dead, and the prince of the kings of the earth. Unto him that loved us, and washed us from our sins in his own blood (Revelation 1:5).

> For the Lord himself shall descend from heaven with a shout, with the voice of the archangel, and with the trump of God: and the dead in Christ shall rise first: Then we which are alive and remain shall be caught up together with them in the clouds, to meet the Lord in the air: and so shall we ever be with the Lord (I Thessalonians 4:16,17).

One of the teachings of premillennialism is that the damned will resurrect at the end of a future thousand years. This appears as false teaching, and can be understood as such in the following Biblical verses:

> Marvel not at this: for the *hour* is coming, in which all that are in the graves shall hear his voice. And shall come forth; they that have done good, unto the resurrection of life; and they that have done evil, unto the resurrection of damnation (John 5:28,29).

The saved and the damned both rise up out of their graves within an hour! Before the rapture occurs, the dead in Christ will resurrect with incorruptible bodies. All who are alive in Christ, on earth, will be turned into dust and then into incorruptible bodies, which is a "change" that will occur "in the twinkling of an eye." Then the rapture occurs! The explanation of the following Biblical verse gives additional understanding about the lost: "And when he had opened the seventh seal, there was silence in heaven about the space of half an hour" (Revelation 8:1). During the half-hour of silence, the incorruptible bodies of the damned will resurrect from their graves. The damned who are alive on earth, along with them, will also receive incorruptible bodies, "in the twinkling of an eye":

> Behold, I shew you a mystery; We shall not all sleep, but we shall all be changed, In a moment, in the twinkling of an eye, at the last trump: for the trumpet shall sound, and the dead shall be raised incorruptible, and we shall be changed. For this corruptible must put on incorruption, and this mortal must put on immortality. So when this corruptible shall have put on incorruption, and this mortal shall have put on immortality, then shall be brought to pass the saying that is written, Death

is swallowed up in victory. O death, where is they sting? O grave, where is thy victory? The sting of death is sin; and the strength of sin is the law. But thanks be to God, which giveth us the victory through our Lord Jesus Christ (I Corinthians 15:51-57).

It is told in the chapter of Matthew 24 about the signs of the end of this dimension of time, space, and matter. Many have wondered if the elect will go through "The Tribulation" that precedes the end. The following verse is from the Old Testament:

And they shall go into the holes of the rocks, and into the caves of the earth, for fear of the Lord, and for the glory of his majesty, when he ariseth to shake terribly the earth (Isaiah 2:19).

The reason for knowing that the elect *will* go through "The Tribulation" is by the following verses: "And woe unto them that are with child, and to them that give suck in those days" (Matthew 24:19)! Two verses down: "For then shall be great tribulation, such as was not since the beginning of the world to this time, no, nor ever shall be" (Matthew 24:21). There is no doubt that these two verses refer to the verse of Isaiah 2:19, because the destruction of Jerusalem by Titus could not measure up to God's wrath when He shakes the earth. The babies of the earth have pure spirits, but will still go through "The Tribulation" like the rest who are alive at that time. However, if we are born-again, we can pray to God the following prayer:

Watch ye therefore, and pray always, that ye may be accounted worthy to escape all these things that shall come to pass, and to stand before the Son of man (Luke 21:36).

Some of the signs that point to Christ's return have occurred during other generations, but this generation has experienced them all and includes one that has never happened before:

When ye therefore shall see the Abomination of Desolation, spoken of by Daniel the prophet, stand in the holy place, (whoso readeth, let him understand) (Matthew 24:15).

On April 14, 1986, the *New York Times* published the first visit of Pope John Paul II to a synagogue in Rome, Italy, which happened on day 13, which was the day before. Synagogues are no longer considered a Holy Place to God in this Dispensation of Grace. However, when Jesus spoke to His disciples in Matthew 24:15, they were under the Dispensation of the Law. The temple and synagogues were Holy Places during that time. When Jesus was speaking of the "Abomination of Desolation standing in the Holy Place," He was projecting His disciples into an unknown future of April 13, 1986. Therefore, what the pope stands for in his organized religion (not the pope himself, necessarily) is an abomination. Jesus will return in the same generation in which all the signs appear: "Verily I say unto you, This generation shall not pass, till all these things be fulfilled" (Matthew 24:34).

The "Abomination of Desolation" is a symbol of "organized religions"—those organized by men and governments. This can be understood by the vision of the prophet Zechariah as told in the entire chapter of Zechariah 5. It tells of the vision of a flying scroll known as "wickedness" and is symbolized as a woman. She was encased in a flying ephah, which was a measuring basket, and covered with a talent of lead. Two women with wings, known as "... the anointed ones that stand by the Lord of the whole earth" (Zechariah 4:14), transported her into the land of Shinar, better known as Babylon. Those two anointed ones are a representation of the "Spirit and Word," and are apt symbols of the power that exposes organized religions. Just as Babylon was a place of literal bondage for the Jews under the Dispensation of the Law, "Mystery Babylon" is a place of spiritual bondage for spiritual Jews under the Dispensation of Grace. Zechariah was told: "... This is their resemblance through all the earth" (Zechariah 5:6). Therefore, while waiting for the Lord's return:

> Do all things without murmurings and disputings: That ye may be blameless and harmless, the *Sons of God* without rebuke, in the midst of a crooked and perverse nation, among whom ye shine as lights in the world (Philippians 2:14,15).

Since time had a beginning, it will also have an end! On the last day, the disintegration of all matter will occur with a cataclysmic implosion in the heavens:

> But the day of the Lord will come as a thief in the night; in the which the heavens shall pass away with a great noise, and the elements shall melt with fervent heat, the earth also and the works that are

therein shall be burned up (II Peter 3:10).

The dust of the universe will fall through an abyss called the "bottomless pit" or "black hole" into the lake of fire:

> For a fire is kindled in mine anger, and shall burn unto the lowest hell, and shall consume the earth with her increase, and set on fire the foundations of the mountains (Deuteronomy 32:22).

God created all things, including Hell! Nowhere in the Genesis story does it explicitly say that God created Hell, yet it had to be that Hell was made possible during the six-day period of creation time. However, there is a place in Scripture that says that God "prepared" Hell:

> Then shall he say also unto them on the left hand, Depart from me, ye cursed, into everlasting fire, *prepared* for the devil and his angels (Matthew 25:41).

It is a fact that Hell was "prepared" from something that God made during the six creative days. On the fourth creative day, God made the sun, moon, and stars. He "prepared" Hell by burning stars that collapsed and caused "black holes," which are abysses that lead into the lake of fire called Gehenna or Hell.

As stated before, when God performed His first creative act called the "Big Bang," the universe began to expand, and is still doing so, because time is still being spent, and time and space are related to each other: Einstein's *General Theory of Relativity* actually bespeaks of a "closed universe"[19] in which the universe will collapse and become a cosmic black hole. The word of God speaks pictorially of a closed universe, in the following verses of Scripture:

> And I beheld when he had opened the sixth seal, and, lo, there was a great earthquake; and the sun became black as sackcloth of hair, and the moon became as blood; And the stars of heaven fell unto the earth, even as a fig tree casteth her untimely figs, when she is shaken of a mighty wind. And the heaven departed as a scroll when it is rolled together; and every mountain and island were moved out of their places. And the kings of the earth, and the great men, and the rich men, and the chief captains, and the mighty men, and every bondman, and every free man, hid themselves in the dens and in the rocks of the mountains; And said to the mountains and rocks, Fall on us, and hide us from the face of him that sitteth on the throne, and from the wrath of the Lamb: For the great day of his wrath is come; and who shall be able to stand (Revelation 6:12-17)?

Cosmologists have stated that there is not enough "black matter" to cause a collapse of the universe. Their theories and calculations, along with the "world without end" premillennialist's teachings cannot standup to Biblical truths.

Notice in particular that verse 14 states: "the heaven departed as a scroll when it is rolled together," which explains how space (firmament or three heavens) will roll backward to the earth. This depicts a closed universe in which all matter (except the glorified bodies of redeemed ones) will end up in the lake of fire. The stars will fall to the earth and not the other way around, as the lower part of Hell is in the lower parts of the earth where Jesus' Spirit descended after He died:

> Now that he ascended, what is it but that he also descended first into the *lower parts of the earth?* He that descended is the same also that ascended up far above all heavens, that he might fill all things (Ephesians 4:9,10).

Many people ignore the warnings that there is a Hell's fire to avoid. The Bible is sometimes looked upon as a book of fables, but as time advances, men of science are discovering facts that prove the Bible true. The black holes of science and the bottomless pit of the Bible synonymously describe the pathways that lead downward into the unquenchable lake of fire. When matter gets too close to a black hole, it is sucked into the hole and heats up as it descends. When the universe collapses, there will be enough matter to heat up and feed the fires of Hell forever, for the heat will neither escape nor become quenched. The glorified Sons of God will have a new heaven and earth in which to dwell forever:

> And I saw a new heaven and a new earth: for the first heaven and the first earth were passed away; and there was no more sea. And I John saw the holy city, new Jerusalem, coming down from God out of heaven, prepared as a bride adorned for her husband. And I heard a great voice out of heaven saying, Behold, the tabernacle of God is with men, and he will dwell with them, and they shall be his people, and God himself shall be with them, and be their God. And God shall wipe away all tears

from their eyes; and there shall be no more death, neither sorrow, nor crying, neither shall there be any more pain: for the former things are passed away. And he that sat upon the throne said, Behold, I make all things new. And he said unto me, Write: for these words are true and faithful. And he said unto me, It is done. I am Alpha and Omega, the beginning and the end. I will give unto him that is athirst of the fountain of the water of life freely. He that overcometh shall inherit all things; and I will be his God, and he shall be my son (Revelation 21:1-7).

God gave us the Bible that we may understand truth and His purpose in that truth. God did not leave us with explicit accounts to confuse us. The writer of Genesis, which tradition attributes to Moses, undoubtedly believed the world would always understand about the marriages of the Sons of God to the daughters of men and the "corruption" they brought upon the earth. God will never again have to destroy the earth with water, because it is impossible for that type of corruption to occur again.

There is no Biblical proof that there was a possible way for people who lived at the time of the great flood to repent and be brought into the ark with Noah and his family. All the people who died in the flood, babies included, were either unjust or imperfect in their generations for having mixed with or descended from the soulless seed of mankind. If Methuselah had lived longer, God would have taken him into the ark, for he was the last one of Noah's "perfect" generations to die before the flood. Nevertheless, Methuselah could not feel deprived of God's blessings, for he undoubtedly lived longer than any other person in the world.

Noah had been a "preacher of righteousness" to the Sons of God, but it was for righteousness sake only that he preached, for the Sons of God could not find a "place of repentance" in those days. If they could have repented, and even if half of them had done so, the ark could not have held them all. Therefore, God did not mean any people other than the eight members of the Noah family to survive the great flood. After God had achieved His purpose in sending the great flood, He left a promise, and a sign of that promise, to Noah and all the rest of earth's life:

> And God said, This is the token of the covenant which I make between me and you and every living creature that is with you, for perpetual generations: I do set my bow in the cloud, and it shall be for a token of a covenant between me and the earth. And it shall come to pass, when I bring a cloud over the earth, that the bow shall be seen in the cloud: And I will remember my covenant, which is between me and you and every living creature of all flesh; and the waters shall no more become a flood to destroy all flesh. And the bow shall be in the cloud; and I will look upon it, that I may remember the everlasting covenant between God and every living creature of all flesh that is upon the earth (Genesis 9:12-16).

After God made Cro-Magnon man in His "image" but *not* in his "likeness," He gave him dominion over the earth. The Sons of God inherited that dominion, for they were also made in the image of God. However, they were not to "replenish the earth, and subdue it" (Genesis 1:28), because they were to live in the

Garden of Eden forever. Since things did not happen as God willed, which was Plan-A, then God gave a new command to the Sons of God after the flood:

> And God blessed Noah and his sons, and said unto them, Be fruitful, and multiply, and replenish the earth. And the fear of you and the dread of you shall be upon every beast of the earth, and upon every fowl of the air, upon all that moveth upon the earth, and upon all the fishes of the sea; into your hand are they delivered. Every moving thing that liveth shall be meat for you; even as the green herb have I given you all the things (Genesis 9:1-3).

God made Adam and Eve in His "image and likeness." Therefore, they were made to have free will like God, Himself. They had the free will to choose to eat of the Tree of the Knowledge of Good and Evil. Something in the fruit is what genetically made them fall! What would have happened had our first parents eaten of the fruit of the Tree of Life first? Knowing that God says that He is a God of life, we realize that eating the fruit of that tree first would have preserved their bodies and souls forever in the Kingdom of Light. God drove them out of the garden, after they fell, so they would not eat of The Tree of Life and be preserved as dust forever in the Kingdom of Darkness. However, if our first parents had not fallen, the Garden of Eden would have continued on forever. The vast amount of space, time, and matter, shows us that God had already made preparation to expand the Garden of Eden to the size of earth, and on into the cosmos—a forever expanding, "open universe." Reproduction of the Sons of God would have continued on forever. Therefore, eternal life with eternal bliss!

We should be thankful that God had Plan-B in mind when Plan-A did not happen!

The Bible conceptually reveals what God created from the beginning of time, to the end of the sixth creative day, which ended about six thousand years ago. When we look at a geological time chart[20] we can see about 2 billion of the 20 billion or more years that God used for the six days of creation. When we consider the amount of time, which makes up the seventh day of rest, we can see that time, all-inclusive, is one cosmic week.

God created all things in a perfect order, by a foreordained plan, but He did not reveal every detail of creation in Genesis. He has left some of the details in the earth and has permitted us to uncover some of them from time to time: "It is the glory of God to conceal a thing: but the honour of kings is to search out a matter" (Proverbs 25:2). As long as mankind lives on the earth, he can help save souls, time, and expense by using the Bible as a source and guide for seeking God and His truth. The Bible declares:

> The earth is the Lord's, and the fullness thereof; the world, and they that dwell therein. For he hath founded it upon the seas, and established it upon the floods (Psalms 24:1,2).

The two verses above declare that Darwin was right when he said that life occurred by evolution, for the first single-celled organisms from which all life descended were "founded upon the seas." Life as we know it today was "established upon the floods" in Noah's day. The flood destroyed not only the "corrupt seed," but also Cro-Magnon man—the "soulless seed," who was made in the "image" of God. Destroying Cro-

Magnon man made it impossible for a "corrupt seed" to ever occur again. An intervention of God with a great flood made creation perfect again. This occurrence could not have been a "random chance" happening for the Bible and the universe loudly declare that:

    God Did It

Dispensations and Times

➤Innocent

>•Not knowing good and evil. When death occurs to the natural bodies of innocent ones, during any period of time, their spirits go right then and dwell with God (Matthew 19:14-Mark 10:14- Luke 18:16).

➤Accountable

>•Knowing good and evil. God deals with accountable people in different ways during different periods of time.

➤Prelaw

>•Sin was dead (Romans 5:13—7:8). At death both Jews and Gentiles went into Abraham's Bosom, also called Sheol, Hades, or Prison (I Peter 3:18,19—4:5,6).

➤Dispensation of Law

>•Sin revived (Romans 7:9). Gentiles not under covenant. At death all went into Abraham's Bosom, etc. Righteous Jews: At death all went into Paradise. A veil separated them from God's Abode. Unrighteous Jews: At death all went into Hell or Gehenna.

➤Transition Period

>•Christ Crucified: His spirit went into Paradise and perfectly purged the Jews. Heavenly veil

became rent. Temple veil became rent. Dispensation of Law ended (Hebrews 9:22,23). Christ's Spirit also went into Abraham's Bosom and preached the Gospel of the Good News to the imprisoned souls. Before resurrecting, Christ also went into Hell and took the Keys of Hell and Death.

➤Dispensation of Grace
- Dispensation of Grace began. Jews first: Pentecost, until end of time. Born-again Jews, at death, go into God's Abode. Gentiles second: The Acts 10, until end of time. Born-again Gentiles, at death, go into God's Abode.

➤Law of Sin and Death
- Accountable Jews, not born-again, at death, go into Hell or Ghenna.

Accountable Gentiles, not born-again, at death, go into Hell or Ghenna.

References

[1] Authorized King James Version of the *Holy Bible,* (New York: The World Publishing Company).

[2] Joseph Laffan Morse (editor in chief), *The Universal Standard Encyclopedia,* (New York: Wilfred Funk, Inc., 1958), 22:7989,7990.

[3] K. Dose, "The Origin of Life," *Grzimek's Encyclopedia of Evolution,* (New York: Van Nostrand Reinhold Co., 1979), 93.

[4] Robert Jastrow, "Genesis Revealed," *Science Digest Special Edition Magazine,* (New York: Heart Corporation, Winter 1979), 34.

[5] D.S. Peters and W.F. Gutmann, "The Meaning of the Theory of Evolution," *Grzimek's Encyclopedia of Evolution,* (New York: Van Nostrand Reinhold Co., 1979), 43

[6] Atlanta (AP), *The Charleston Gazette,* 28 July 1979.

[7] Richard E. Dickerson, "Chemical Evolution and the Origin of Life," *Scientific American Magazine,* 239,3 (Sept. 1978): 70.

[8] K. Dose, "The Origin of Life," *Grzimek's Encyclopedia of Evolution,* (New York: Van Nostrand Reinhold Co., 1976), 107.

[9] W. Herre and M. Rohrs, "The Evolution of Domestic Animals." *Grzimek's Encyclopedia of Evolution,* (New York: Van Nostrand Reinhold Co., 1976), 520.

[10] Ibid., 529.

[11] Adam Clarke, "Genesis-Deuteronomy," *Clarke's Commentary,* I (New York, Cincinnati: The Methodist Book Concern,): 63.

[12] Max R. Gaulke, "Kingdom Foregleams," *May Thy Kingdom Come...Now,* (Anderson, IN: Warner Press, Inc., Publication Board of the Church of God, 1959), 19.

¹³Adam Clarke, "Genesis-Deuteronomy," *Clarke's Commentary,* I (New York, Cincinnati: The Methodist Book Concern): 68.

¹⁴Charles F. Kraft, "The People of God," *Genesis, Beginnings of the Biblical Drama,* (Woman's Division of Christian Service Board of Missions, The Methodist Church, 1964), 140.

¹⁵E. Leslie Carlson, "Money, Weights, and Measures," *Bible Study Helps of the Holy Bible,* (New York: The World Publishing Company), 29.

¹⁶"In Search of Noah's Ark," *Christian Life Magazine,* (1977).

¹⁷Max R. Gaulke, "Kingdom Foregleams," *May Thy Kingdom Come...Now,* (Anderson, IN: Warner Press, Inc., Publication Board of the Church of God, 1959), 22.

¹⁸Adam Clarke, "Matthew-Acts," *Clarke's Commentary,* V (New York, Cincinnati: The Methodist Book Concern): 684.

¹⁹Michael Zeilik, "Einstein and the Unity of the Universe," *Astronomy, The Evolving Universe,* (Harper & Row Publishers, 1976), 127.

²⁰David B. Guralnik, *Webster's Dictionary: New World of the American Language,* (Nashville, TN: The Southwestern Co.) "Geological Time Chart" 314.